EU beef farming systems and CAP regulations

EU beef farming systems and CAP regulations

Editors:
Patrick Sarzeaud
Andie Dimitriadou
Milan Zjalic

EAAP Technical Series No. 9

ISBN 978-90-8686-058-6
ISSN 1570-7318

First published, 2008

Wageningen Academic Publishers
The-Netherlands, 2008

Table of contents

Introduction and acknowledgments

With a production of 7,8 million tons in equivalent carcass, the European Union (EU-25) holds, in 2006, the third world rank of beef production, second only to the United States and Brazil (Livestock Institut, 2007). Of this beef production, approximately 60% comes from dairy herds, numbering 22,8 million dairy cows, and 40% from beef herds, resulting from the nursing activity of 12,2 million suckling cows. The former are mainly located in the northern part of the EU, while the latter in France, Spain, Ireland, United Kingdom and Italy.
Since 1995, the EU-15 recorded a fall in its production of approximately 11%, and the enlargement to 25 members did not slow down the decrease. Actually, this process is mainly due to the negative impact of the important fall in the number of dairy cows (-18%), in relation to the growth of milk productivity in a context of fixed quotas. This decrease has not been compensated by the development of suckling cow herds, neither by the increase in carcass weight.
The BSE crises shocked the beef market in 1996 and 2001 and in both cases the EU consumption dropped by 5%. However, by means of special sanitary regulations and the reorganisation of the market through national products, consumption came back to the previous level of 18 kg beef and veal per capita and per annum, nevertheless far from the 42 kg level in the USA and 33 kg in Brazil. Therefore, with the decrease in production, EU beef imports are now higher than exports and contribute to boosting up the competition between world meat producers.
The International Farm Comparison Network, led by the FAL (Federal Agricultural Research Centre) in Braunschweig (Germany) and in charge of the analysis of cost production through worldwide farm data, demonstrated what part of the CAP premiums is engaged in the sustainability of EU beef farming systems. For the majority of the European beef producers, the costs of production are higher than the market price (Agribenchmark report, 2006). That is why any changes in common policy or any WTO resolutions are becoming more and more relevant for the future of beef farming systems.
The new CAP applied by EU countries since 2005 appears thus as a big change, with the implementation of the Single Payment Scheme (SPS) and the decoupling of subsidies from production. According to the prevailing regulations, beef production in Europe is amongst the most affected sectors. Therefore, the question is: what could be the impact on the future of European beef farming systems and how could beef farmers adapt in order to improve the sustainability of their farms?

The Beef Task Force: a new forum for the analysis of policy impact on beef farming systems

Based on the above analysis, the European Association for Animal Production (EAAP) established the Beef Task Force, a group for the exchange of information within the EAAP Cattle Network Working Group, with the objective to monitor developments in the European beef sector, to set up policy impact analyses on production and farm strategies, and to identify research needs in that sector.
The Beef Task Force consists in a network of correspondents and researchers of EU member countries (Italy, Spain, France, Germany, Ireland and Sweden), working in Ministries and research centres on economics and animal husbandry. The above countries account for approximately 80% of the European beef and veal production. The coordination has been assigned to the French Livestock Institute. The BTF could be enlarged to include other members from other countries, according to its research needs. Beyond the task of exchanging information and experience, the Beef Task Force aims also to be a body for the development and coordination of Community-financed research projects, while its recommendations will be forwarded to the EAAP Council, to EAAP Member Organizations and to national and EU representative farmers' associations.
The partners' expertise lies in the observation of beef farming systems (BFS) through farm networks and economic analyses and simulations. After a first phase of methodological coordination (2005),

the Beef Task Force engaged in the analysis of the implementation of the CAP reform (Agreement of Luxembourg 2003 - 'Mid Term Review') and its consequences on BFS.
Four parts have been jointly produced in order to create a common analysis:

- Proposal for a classification of beef farming systems: description, location and quantification of each type, as a basis of any economic study with a test on the FADN-RICA (Farm Accountancy Data Network) database.
- Analysis of the different ways of CAP implementation among European countries.
- Country analyses of the – real and expected – consequences for beef farming systems.
- The synthesis of perceptions about the future of European beef farms, their contribution to the market and to land use.

The group presented its report at the 3rd EAAP - Cattle Network Workshop '*Profitability and sustainability of beef farming: Adaptation and conformation of EU beef systems to CAP regulations*' held on 24 August 2007 in Dublin, Ireland, in the context of the 58th Annual Meeting of EAAP. The Workshop was sponsored by BRAUNVIEH AUSTRIA – Federation of the Austrian Brown Swiss Cattle Breeders.
The conclusion of this report and the organisation and successful outcome of the Workshop in Dublin is a result of an efficient collaboration between partners whom we wish to thank sincerely, as well as the members of the EAAP Cattle Network who gave us the opportunity to engage in this project.
Members of the Beef Task Force who collaborated for this report are:

- Patrick Sarzeaud, Frédéric Bécherel (French Livestock Institute, France)
- Claus Deblitz, Daniel Brüggemann (Institute of Farm Economics, Germany)
- Kees de Roest, Claudio Montanari (Research Center for Animal Production, Italy)
- Gerry Keane (Teagasc, Ireland)
- Ana Redondo, Ernesto Abel Reyes (MAPA, Tragsega, Spain)
- Pernilla Salevid, Theres Strand (Taurus, LRF Konsult, Sweden)
- Andie Dimitriadou, Milan Zjalic (EAAP)

An update on the CAP Reform and the beef sector in the European Union

M. Zjalic, A. Rosati and A. Dimitriadou

European Association for Animal Production, Via G. Tomassetti 3, 00161, Rome, Italy

Abstract

The present paper contains parts of the article '*Beef Production in the European Union and the CAP Reform*', prepared by the authors in January 2006 and published on the EAAP website, in addition to some updates on developments in the last two years. The article has been translated and published in professional and scientific journals in several countries (Hungary, Croatia, Poland, Czech Republic, and others).

Keywords: CAP reform, beef production, beef consumption, beef trade, future perspectives.

Basic facts

Production

With a total production of some eight million tons, the EU represents about 13% of the total production of beef and veal in the world. In the last decade, beef production in the EU-15 has been gradually declining with a total drop of 7.6% in 2003 compared with 1993 (Figure 1). In 2004, beef production in the enlarged European Union (twenty-five Member States) equaled 1993 production in the EU-15. Production in other European countries also declined: in 2004 Europe produced 11,6 million tons of beef compared with 14,8 t in 1993-95. During the same period, the global beef production increased by 6,3 million tons, or by 16,8% due to the substantial increase in other continents, particularly in South America (increase of 2,6 million tons or 26%).

Following the high production of 2004, due to increased end-of-year slaughtering in the Member States that started applying decoupling from 2005, EU-25 net production declined considerably in

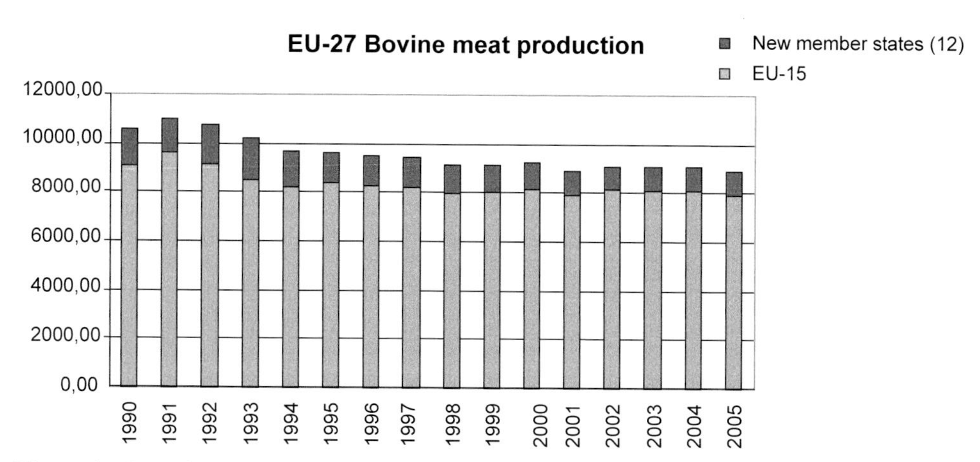

Figure 1. EU-27 bovine meat production 1990-2005 (1000 tonnes).

2005, falling by 2.5%. The EU-25 beef and veal production is estimated to have increased in 2006 to 8.0 million tons, up by 2%. This increase is mainly linked to the impact of the end of the Over Thirty Months Scheme (OTMS) in the UK with more beef entering the market.

Consumption

In several previous decades, the marketing of beef has been confronted with some negative connotations associated with red meat consumption (e.g. cholesterophobia). In addition, the BSE crisis and the impact of the outbreak of foot-and-mouth disease in the nineties heavily affected beef consumption: in fact, beef and veal consumption per head increased from 18.2 kg/head in 1960 to almost 25 kg in 1985, and has since then declined again to 19.3 kg/head in 2000. However, beef consumption experienced a strong recovery in 2002, which was also confirmed in 2003. The European Union and its Member Countries reacted promptly by introducing a system of improved control of animal health, ban of slaughter of animals under risk, control of movements of live animals and meat, registration and identification of all cattle, as well the support of voluntary introduction of labelling of beef.

In the period 1995-2005, beef consumption fell dramatically in the new Member States (up to 50%), in line with the strong reduction in beef production. Low consumers' preferences for beef and the consumption of veal are also among the reasons for the great difference between new and old Member States. The new 10 Members produce 8% of the total beef and veal and consume 6% of the beef and veal consumption in the EU-25.

The 2001-2006 period was marked by a recovery of beef consumption (Figure 2) and its stabilisation at the level of some 20 kg per year per head in the western and some 7 kg in the central and eastern parts of the European Union. A programme financed by the European Commission, aimed at informing consumers on the measures undertaken to ensure safety and quality of beef, greatly contributed to the recovery of beef consumption. However, the substantial increase in beef consumption was caused by the consumers' preference for beef and other meats compared to poultry due to the effect of the outbreaks of Avian Influenza, the re-entry of 'Over Thirty Months' beef on the UK market and the re-opening of the single market for British beef and live animal exports.

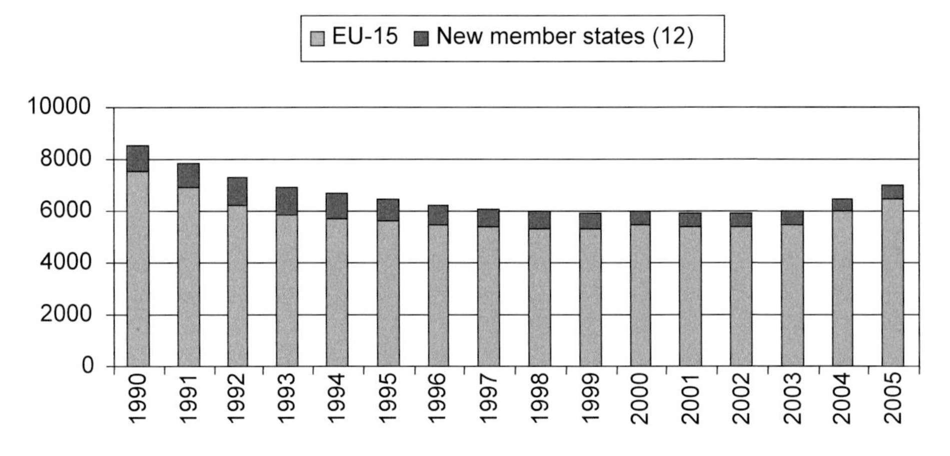

Figure 2. EU-27 bovine meat consumption 1990-2005 (1000 tonnes).

Trade

Historically, the European Union has been a major beef exporter. During the period 1999-2004, the annual export of beef averaged 597 170 tons (varying from 973 000 tons in 1999 to 397 000 tons in 2004: a decrease of 59%). The import of beef varied from 423 000 tons in 1999 to 608 000 tons in 2004: an increase of 44%. During the same period, new Member States exported on average 144 170 tons of beef annually (varying from 121 000 tons in 1999 to 188 000 tons in 2004: an increase of 55.2%) and imported 44 500 tons (with an increase of 171.5% between 1999 and 2004). A declining trend in export was due mainly to reduced production. Decline in beef production was caused mainly by the outbreak of diseases (BSE crisis, FMD) and consequent decline in consumption and demand for beef. In the same period, the beef sector in countries of the Central and Eastern Europe completed adjustments to the market conditions (sharp decline in domestic consumption, loss of export markets in the former USSR), which resulted in 30 to 50% lower production compared with the pre-transition period. Some policy changes introduced in the EU in line with agreements reached within the World Trade Organisation, such as reduced expenditures for domestic market support, decrease in export refunds and lower border protection, also contributed to the decrease of exports and increase of imports of beef. Decrease in supply of calves from dairy herds also affected beef production: in 2004, the number of dairy cows was by 30% lower than in 1990 as a result of milk quotas and the constant increase in per cow milk production. Since 2003, EU is a net importer (Table 1). The recovery of consumption also contributed to the decline of export surpluses.
Extra EU-25 meat exports fell by 33.5% in 2005 as a combined effect of a strong euro, relatively high internal prices, cuts in export refunds and lower net production. Extra EU-25 meat exports will continue to be constrained by low domestic availability and lower competitiveness and are thus projected to decline further.

Table 1. EU-25 Supply balance beef/veal 1000 tonnes (carcass weight).

	2002 (EU-15)	2003 (EU-15)	2004	2005
Gross internal production	7,502	7,387	8,135	7,910
Net production	7,467	7,358	8,041	7,841
Change in stocks	-55	-204	-34	0
Imports (total trade, except live animals)	445	470	504	520
Exports (total trade, except live animals)	472	373	328	218
Intra-EU trade (all trade, incl. live animals)	2,138	2,434	2,646	2,661
Internal use (total)	7,495	7,660	8,251	8,143
Gross consumption (kg/head/year)	19.9	20.2	18.0	17.7
Self-sufficiency (%)	99.6	96.1	97.5	96.3

Source: Agriculture in the European Union - Statistical and economic information 2006

The CAP reform 2003 and the beef sector

The latest reform of the Common Agricultural Policy, agreed in June 2003, introduced the Single Payment Scheme (SPS), a system of annual aid paid to producers irrespective of production ('decoupled'). The SPS combines a number of existing direct payments received by farmers in a single payment, determined on the basis of payments received over a reference period. The June 2003 agreement established a maximum amount each State could use for direct aid payments - known as the national ceiling - based on the total of direct aids paid over a reference period in each Member

State. Ceilings are established within the overall national ceilings for each of the main products for which aid was paid in the past.
Member States may opt to introduce the SPS in full, combining all aid in one payment or decide to maintain a proportion of direct aids to farmers in their existing form ('partial decoupling'), mainly where they believe there may be disturbance to agricultural markets or abandonment of production as a result of the move to the SPS.
Following the reform, beef payments are therefore either incorporated into the SPS, or they may be paid as follows in partially decoupled payments:
Member States can opt for keeping up to 100% of the 'suckler cow premium' and up to 40% of the 'slaughter premium for adult bovine animals' coupled. Alternatively, they may keep 100% of the slaughter premium for adult bovine animals coupled or, instead, up to 75% of the 'special male premium'. And finally, Member States may retain up to 100% of the 'calf slaughter premium' amounts to be made as a product-specific payment.
The premium rates for bovine animals (before application of the partial decoupling percentages) are shown in Table 2.

Table 2. Premium rates for bovine animals.

Premium	Rate
Suckler cow premium	200 € / animal
Slaughter premium	80 € / animal
Calf slaughter premium	50 € / animal
Special male premium	150 or 210 € / animal (depending on type)

The extensification payment (currently ranging from €40 to €100 per animal according to the density chosen by the Member State), additional payments and the deseasonalisation premium will be fully decoupled.
Where Member States decide to maintain product-specific aid payments, limits on those payments apply, via the national ceilings on the amount of aid available and limits on the number of animals which can generate entitlement to aid payments. There are also individual ceilings corresponding to the number of premium entitlements allocated to each farmer in the case of suckler cow premiums.
The single payment scheme came into operation on 1 January 2005. Member States had the option to apply the SPS after a transitional period (until 31 December 2005, or 31 December 2006), where special agricultural conditions so justified. Where the transitional option is taken, Member States must apply the 2004 rules relating to livestock aid payments – for example, limits on numbers of livestock kept per hectare of fodder area (stocking density) – and an extensification payment.
Public storage of beef plays a role in supporting the beef market. The EU now buys beef into 'intervention' stores only when average market prices in a Member State or a region of a Member State fall to below €1560/ton over two consecutive weeks. There is also a system of Private Storage Aid (PSA) under which private traders are encouraged, by means of a partial subsidy, to store beef temporarily at times of oversupply. Intervention is thus now regarded as a 'safety net' providing market support limited in time rather than a market management tool, and PSA is the preferred option to deal with temporary oversupply.
Member States may grant 'additional payments' to support agricultural activities that are important for the protection or enhancement of the environment or for improving the quality and marketing of agricultural productions, including the livestock sector. These 'additional payments' may use up to 10% of the funds that are available for a certain sector included in the single payment scheme

in the Member State concerned. The additional payments must be within the overall ceilings laid down for the sector in question.

National/regional flexibility

Member States have various options in the way they calculate and make payments: they may calculate SPS on the basis of individual farmers' direct payments during a past reference period, or average all past payments and paid uniformly over a region or state. Member States may either apply a mixed historic/flat rate approach that stays the same over time ('static'); or they may choose a mix that alters over time ('dynamic'), usually so that the proportion of SPS based on historic references reduces as the flat rate element increases, offering a means to transit from the basic to the flat rate approach.
In the first years after accession, the new members may opt for a different type of direct aid scheme: the 'Single Area Payment Scheme'. However, if they apply the SPS, then the same rules apply as elsewhere in the EU (using the phasing in rates agreed in the accession treaties).

Cross compliance and good environmental practices

Farmers must maintain their land in good agricultural and environmental conditions and respect other 'cross compliance' standards in order to receive SPS and other direct payments. Member States define minimum requirements for good agricultural and environmental conditions. There are also statutory management requirements, set up in accordance with 19 EU Directives and regulations relating to the protection of the environment including animal health and animal welfare. Livestock producers must comply with these rules if they wish to receive direct payments (SPS and/or partially coupled payments).

Maintenance of permanent pastures

Member Sates are obliged to ensure that the area of permanent pasture does not decrease as a result of reforms. In case the area of permanent pastures decreases significantly, the concerned Member State may introduce measures at farm level to oblige farmers to maintain the share of permanent pasture on their holdings.

Perspectives and projections to 2013

The implementation of the single farm payment scheme as part of the Common Agricultural Policy (CAP) reform allows Member States to choose among different options, which will influence the degree of 'decoupling' of the payments. Member States have communicated their preferred option and, based on this information, it has been estimated that in 2013 approximately 90% of the budgetary transfers in the form of direct payments (including national envelopes and top-ups) for the arable crops, milk, beef and sheep sectors will be part of the single farm payment for the EU-25 as a whole. The rate would be higher for the milk (100%) and arable crop (93%) sectors than for beef and sheep sectors (78% and 73% respectively).
Overall EU-25 beef production is expected to decrease over the medium term to slightly below 7.5 mio t in 2013, a reduction of 5% from 2005 (Figure 3).
A slight increase of consumption due to higher availabilities in 2006 is expected to be followed by a relative stagnation over the medium term. In the new Member States the potential increase – fuelled by rising income levels – would be broadly offset by the sustained price increase and the low consumer preference for beef.

A relatively steady demand and tight domestic supply are expected to result in firm prices over the projection period, attracting more imports of beef entering at full duty, notably high-quality beef cuts from South America. Following a short-term setback in 2006 due to import restrictions imposed on Brazil as a consequence of FMD, total beef imports are expected to resume their growth and exceed 0.7 mio t by the end of the projection period.

Higher profitability of EU-25 production due to higher domestic prices and the abolition of export refunds for live animals for slaughter led to a considerable decline of live animal exports in 2005 (-34%) that are projected to remain at a low level throughout the forecast period.

With around two thirds of European beef production originating directly or indirectly from the dairy cow herd, the medium-term development of beef production would continue to be highly dependent on developments in the milk sector. Given the increasing trend in milk yields, the dairy cow herd would continue to decrease with the effect of lower calf availability and hence a lower level of fattening activities.

On the other hand, the suckler cow herd would increase mainly in those countries which keep most of the animal premia coupled. However, this development is not expected to counterbalance the effects resulting from both the decreasing EU-25 dairy cow herd and shrinking suckler cow herds in Member States which have fully decoupled and in certain cases additionally re-distributed direct payments.

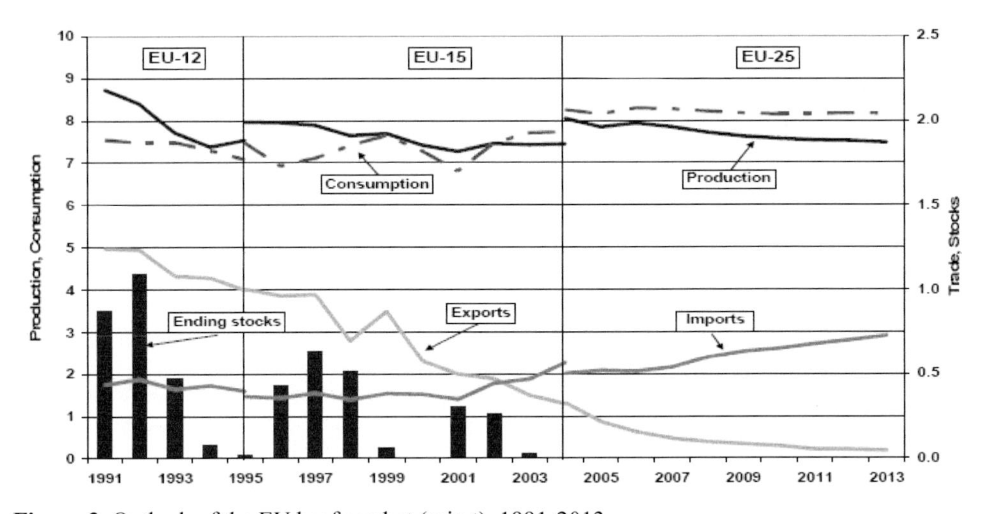

Figure 3. Outlook of the EU beef market (mio t), 1991-2013.

References

Commission Regulation (EC) No 795 of 21 April 2004 (OJ L 141, 30.04.2004).

Commission Regulation (EC) No 796 of 21 April 2004 (OJ L 141, 30.04.2004).

Commission Regulation (EC) No 1973 of 29 October 2004 (OJ L 345, 20.11.2004).

Council Regulation (EC) No 1782 of 29 September 2003 (OJ L 270, 21.10.2003).

Cunningham, E.P. and the European Association for Animal Production (Eds), 2003. After BSE – A future for the European livestock sector, Wageningen Academic Publishers, EAAP publication no 108.

European Commission, Directorate-General for Agriculture, 2006. Prospects for Agricultural markets and income in the European Union 2006-2013. http://ec.europa.eu/agriculture/publi/caprep/prospects2006b/index_en.htm.

European Commission, Directorate-General for Agriculture, 2003. Reform of the Common Agricultural Policy. Medium-term prospects for agricultural markets and income in the European Union 2003-2010 http://europa.eu.int/comm/agriculture/publi/caprep/prospects2003b/fullrep.pdf.

European Commission, Agricultural trade statistics: http://europa.eu.int/comm/agriculture/agrista/tradestats/index_en.htm.

European Commission, Agriculture: http://europa.eu.int/comm/agriculture/index_en.htm

European Commission, CAP reform http://europa.eu.int/comm/agriculture/capreform/index_en.htm.

FAOSTAT – FAO Statistical Databases 2007 http://faostat.fao.org.

Mid-Term Review: the implementation of the 2003 CAP reform in the EU

C. Deblitz

Asian Agribusiness Research Centre, Charles Sturt University, Orange, NSW 2800, Australia; claus.deblitz@fal.de

Abstract

Since 2005, the 2003 Mid-Term Review of the CAP has been in progress. Its implementation has been relatively different among the European countries, according to their national choices regarding premium decoupling. If two thirds of the beef production in Europe are fully decoupled, payments remain coupled for two thirds of the EU suckler-cows.

Keywords: mid-term review, CAP implementation, decoupling

CAP evolution: from food sufficiency to environmental, food safety and animal welfare standards

Since the beginning, the EU cattle production has been significantly led by the Common Agricultural Policy. At first, the CAP was based on precise key objectives, notably the need to guarantee self-sufficiency in basic foodstuffs. During the 90's, unfortunate side effects began to emerge, such as the beef and cereal surplus, trade distortions on the world market and concerns about the environmental impact of the CAP.

The above developments necessitated a new orientation of the CAP, planned through the 'Agenda 2000' reform: the policy of subsidiarity links production to the respect of environmental, food safety and animal welfare standards. The second pillar, a special fund engaged to support those standards, has been promoted with the extension of a global subsidy modulation.

Following the main disease crises (1996-2001), and in order to recover consumers' confidence, this orientation has been reinforced with the so-called Mid-Term Review (Luxembourg 2003) that gave even more room to subsidiarity and proposed the decoupling of premiums from production, with the aim to support not only farming, but the long-term livelihood of EU rural areas as a whole.

Finally, the debate on the sustainability of the CAP budget led the European Commission to first simplify the premium mechanism in a single farm payment, and second to propose a reduction in the budget after 2013.

Mid-Term Review 2003: the main decisions

The CAP-reform of 2003, also referred to as Mid-Term Review (MTR), constitutes a major change in agricultural policy. The main characteristics of the MTR are:

- The decoupling of direct payments from actual production (beef, suckler-cows, cereals, milk, etc.). This means that producers receive payments even if they do not produce anymore. Furthermore, payments may not appear as receipts in the beef or cow-calf enterprise anymore, that is receipts – and profitability – are reduced by the amount of previously coupled payments. The level of payments is basically based on the annual average of the historic payments received in the years 2000-2002.

- The linkage of the payments to the fulfilment of regulations regarding the maintenance and management of the land and environment (cross compliance). If a recipient of payments does not comply with the regulations, payments may be cut or even withdrawn.
- Modulation of the payments. Within the implementation period it is possible to reduce the now-decoupled payments and redirect them into the so-called 'second pillar' policy measures, mainly consisting of agri-environmental measures and rural/structural policy measures.

Co-existence of different ways of implementation

The above general principles have been modified in many countries of the EU. The result is a co-existence of different ways of implementation, which can be summarised as follows (only those related to beef and cow-calf production):
- Some countries have opted for keeping some direct payments partially or fully coupled to the animals maintained or produced. Apart from the full decoupling of payments, member states could choose between three options shown in Figure 1.
- The basis for paying the decoupled payments varies between a) a 'single farm payment' (SFP, exclusively based on the farm-individual historic payments), b) an acreage-based payment (homogeneous payment for all land in one region, mainly in the new member states) or c) a hybrid model which is a mixture of the previous two payment systems. The hybrid (or combi) model can be static (relation between acreage payment and SFP remains constant over time) or dynamic (the SFP is phased out in favour of the acreage payment).
- The start year of the implementation was 2005 for all countries with the exception of France, Spain, Netherlands, Finland and Greece (2006).

Figure 2 shows the different payment models and the start year of implementation of the MTR.
- Most countries in southern Europe chose the Single Farm Payment exclusively based on the farm-individual historic payments in order to preserve the amount of premium per farm and to simplify the mechanism. It is the case of Ireland, France, Italy, Spain and Greece.
- Eastern EU countries were oriented to the homogeneous payment for all land in one region, that is acreage-based payment, because of their lack of payments history.
- Finally, some countries such as Germany, Great Britain and northern countries opted for the hybrid model.

Payment	Option I	Option II	Option III
Slaughter-premium **calves**	up to 100 %	up to 100 %	up to 100 %
Suckler-cow premium	up to 100 %	0 %	0 %
Slaughter-premium **adult cattle**	up to 40 %	up to 100 %	0 %
Special premium for **male cattle**	0 %	0 %	up to 75 %

Figure 1. Options for (de)coupling of beef payments in the beef sector (EU-COM, 2005).
Note: Percentages indicate the level of coupling

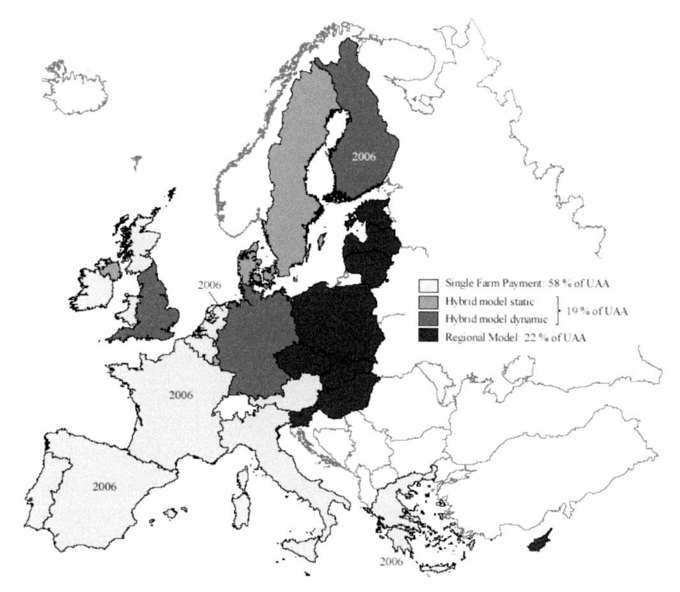

Figure 2. Payment models and start of the implementation of the CAP-reform in the EU-25 (EU-COM, 2005).

The case of Germany, England and Finland – dynamic hybrid model: towards homogeneity of payments

Germany, England and Finland opted for a dynamic hybrid model which is explained here taking Germany as an example. Implementation in the other countries has the same principles but with country-specific differences. Germany opted for full decoupling of the direct payments and a dynamic hybrid (combi) model as follows:

- For each farmer, a part of the decoupled payments is paid as a SFP based on the historic payments he received, while another part is paid on an acreage basis. Figure 1 in the German part of this Report shows which part of the formerly coupled payments were converted into SFP, arable and (as a newly introduced payment) grassland payments, respectively.

The acreage payments are homogenous for defined regions (Bundesländer) with slight differences between the regions. Further, in the beginning of the implementation, there are different acreage payments for cropland and grassland (for regionalised acreage payments see Table 1 in the German part of the Report).

From 2005 to 2009, the relationship between the SFP and the acreage payments will remain constant. In the four-year period 2010 to 2013 the following two main changes are going to occur in the system:

1. The SFP will gradually be phased out (to zero) in four steps in favour of the acreage payments.
2. The grassland payments will gradually increase until they reach the level of the cropland payments.

As a consequence, in 2013 there will only be homogeneous acreage payments for cropland and grassland while slight differences between the Bundesländer will remain. The initial cropland premium may increase or decrease in that period depending on a) the share of SFP in the total payment amount and b) the share of grassland in the Bundesland.

Beef production partially decoupled

The pie-charts of Figure 3 contrast the share of each country in the total EU-25 beef production and in the total suckler-cow numbers with the (de)coupling models applied in each country.
As Figure 3 suggests, approximately two thirds of the beef production in Europe is fully decoupled. The majority of the remaining one third continues to receive 40 percent of the slaughter premium as a coupled premium, with all other payments decoupled. Taking into account the proportion that the slaughter premium has had in total of the previously fully coupled situation, this means that in average only 13 percent (bulls), 8 percent (steers) and 40 percent (heifers) of the previous payment levels remain. It can be assumed that, at least for male animals, the low levels of payments remaining coupled will not lead to different decisions about continuing or stopping production when compared with the fully decoupled situation.
Figure 3 also shows that the situation of the cow-calf production is somewhat different. Not taking into account the share of slaughter premiums (rather unimportant for a suckler-cow herd), payments for two thirds of the suckler-cows on EU-level remain fully coupled, mainly because the two dominant suckler-cow countries in Europe, France and Spain, have opted for this. Contrary to the slaughter premiums, the suckler-cow premium is between €180-200 per cow, a level that will most likely have a significant impact on the decision whether to continue cow-calf production or not.

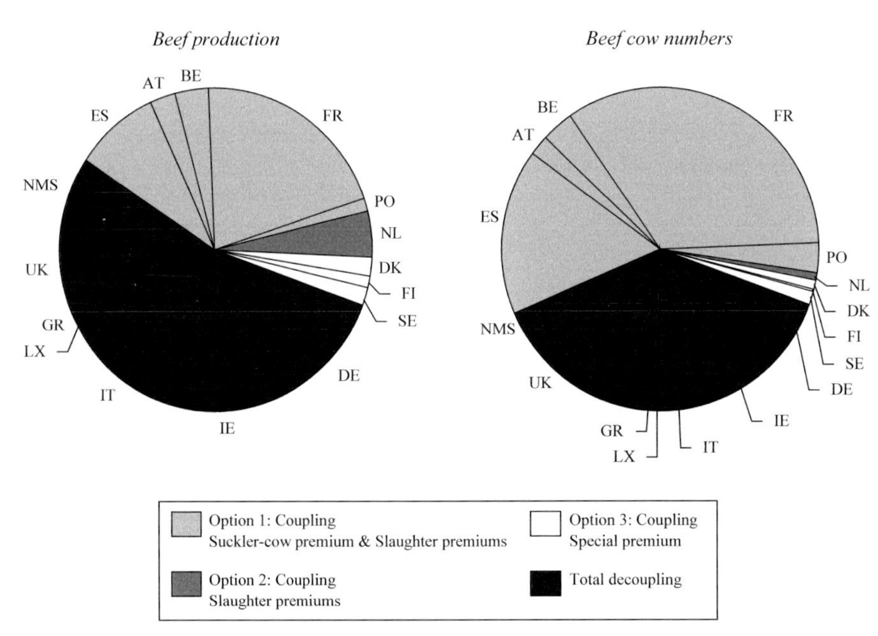

Figure 3. Beef production and suckler-cow numbers in the EU-25 and (de)coupling models by country (EUROSTAT, agra-europe, 2004).

A classification of European beef farming systems

P. Sarzeaud[1], F. Bécherel[2] and C. Perrot[3]

[1]*Institut de l'Elevage, Rond point Le Lannou, 35042 Rennes cedex, France*
[2]*Institut de l'Elevage, Bd des Arcades, 87060 Limoges Cedex 2, France*
[3]*Institut de l'Elevage, 149 Rue de Bercy 75595 Paris cedex 12, France*

Abstract

With the objective of defining standard procedures for monitoring and analyzing the economic efficiency of beef farm operations, the Beef Task Force of the EAAP Cattle Network Working Group developed a classification of the main EU Beef Farming Systems (BFS) – combining data on type of farming, beef activity and an EU zoning. A first implementation of this classification on 2004 FADN (Farm Accountancy Data Network) data gives a good overview of the involvement of each BFS in beef production and their main characteristics.

Keywords: FADN, classification, beef farming systems

Note on the EAAP Cattle Network's Beef Task Force

One of the objectives of the Beef Task Force (BTF) of the Cattle Network Working Group of EAAP is to develop tools and methods for policy impact analysis and for monitoring and understanding farm strategies. The first step undertaken by this group of experts was to establish a classification of the European beef farming systems (BFS) in order to illustrate the diversity of the farms, identifying their localization on the territory and their structural evolution. As an introduction to the national analyses of the impact of the CAP on beef farming systems, this paper presents the classification of beef farming systems in the EU and some illustrations utilizing FADN data.

The classification: a tool for BFS analysis

The improvement of the typology of farming systems is one of the main methodological objectives of farming systems' analysis. From the observation of livestock management through farm panels or networks, differences can be detected between farms in the combination of enterprises, land management or husbandry practices (feeding or breeding). As shown in many studies (see the following national parts), those differences in management and production are significant for the sustainability of farms and their development. The relevance of those differences leads economists to classify enterprises into homogeneous groups, keeping structural and functional components of the systems as the main criteria for classification. This typology is useful:

- for understanding and dealing with the diversity of farms;
- for describing the main characteristics of each system and their importance in the whole population;
- for illustrating the effect of the systems on land use and production;
- for proposing representative systems for economic analysis.

Other methods can also help to improve classification, like statistical analysis (for example, principal components analysis). The objective of this pragmatic approach is to link the sustainability of the systems to criteria that are easy to collect or retrieve in existing databases.

Proposal for a classification of beef farming systems

With respect to beef production, three main classification criteria need to be pointed out:

- The type of farming.
  This means the degree of specialisation of the farm. Beef production is often managed as a complement to other activities: dairy, sheep or crop production. In order to identify as much as possible the beef producers and the farming systems, it is necessary to distinguish this activity on the basis of its economic importance and its contribution to the production and the land use.
- The type of cattle enterprise.
  In this paper, different enterprises and their main outputs are defined as follows:
  - Dairy (or milk) enterprise: Milk production, culling of animals, replacement heifers, breeding animals, surplus calves (usually 1-4 weeks old). The dairy enterprise consists in a continuous, year round activity.
  - Cow-calf (or suckler-cow) enterprise: Weaner calf production, culling of animals, replacement heifers and breeding animals. The cow-calf enterprise ends with the weaning of the calves.
  - Backgrounding (or store cattle) enterprise: Production of pre-products for final finishing, starting from dairy calves or weaners.
  - Finishing (or fattening) enterprise: Production of the final product by taking animals to their final weight: bulls, steers, heifers, cows, calves based on calves from dairy, weaners from cow-calf and backgrounders.
- The location.
  The location of each system is relevant to the agronomic potential of the area. The proportion of land with good potential to produce grass or crops will explain the ability of farmers to produce different types of beef for finishing. Some regions are particularly well equipped for beef production, with a good network of beef cooperatives and abattoirs.

Other criteria available in the FADN database will be used to describe farming systems and their management: e.g. acreage, proportion of grass, stocking rate, and labour units.

Classification based on FADN data

This study follows the work of Vincent Chatellier (see References), who carried out a first analysis of the FADN data, a method of reference to define the categories of beef farming systems. The evaluation of the beef extensification payment commissioned by DG AGRI in 2007[1], gave the opportunity to test an extended classification of beef farming systems proposed by the Beef Task Force. Such classification includes the overall perspectives of the beef farming systems, including the dairy farms engaged in beef production and the differentiation of cow-calf production and finishing enterprises.

FADN data consist in agricultural statistics collected by the Directorate-General Agriculture in an annual harmonized survey. The database contains a collection of descriptive data on a sample of 74,000 commercial farms in the EU-25 (2004), on crops and livestock production. A commercial farm is defined as a farm which is large enough to provide a main activity for the farmer and a level of income sufficient to support his or her family. In order to be classified as commercial, a farm must exceed a minimum economic size. This size is expressed in European Size Units (ESU), evaluated on the calculation of the Standard Gross Margin (SGM), and is qualified in each country.

[1] The article is based on the methodology and intermediate results of the evaluation of the beef extensification payment commissioned by DG AGRI. The authors (from the Institut de l'Elevage) are members of the consultants' team for this evaluation.

Non-commercial farms represent 19% of the enterprises and 0.5% of the Standard Gross Margin for the EU-15.
This panel allows the association of structural, technical and economic parameters in a classification of farming systems. Its large size covers the diversity of the systems and enables extrapolation to the whole population on request, for example at country level.
The beef enterprises can be combined into farming systems as shown in Table 1.
This classification uses criteria available through the FADN data and combines :
- the type of farming, defined in terms of the relative importance of production through their contribution in the Standard Gross Margin
- the beef activity, according to the presence of animals (sex, age, dairy/suckler cows, etc).

Table 1. Criteria that differentiate beef farms.

Base	Production	Farming systems	Criteria
FADN farms > 1 LU All types of farms	Small farms		< 5 dairy cows, < 5 suckler cows, < 8 LU
	Dairy farms		> 5 dairy cows
		Dairy and beef	(With males, with or without suckler cows) > 0.2 male (> 1 year old)/Dairy cow or > 5 suckler cows
		Dairy pure	Others > 5 dairy cows, < 5 suckler cows
	Cow-calf farms		> 5 suckler cows, < 5 dairy cows, LU/(SC+DC)<8
		Cow-calf pure	< 0.2 male (1 to 2 years old)/Suckler cow
		Cow-calf and bull	> 0.2 male (1 to 2 years old)/Suckler cow
		Cow-calf and sheep (goat)	> 20% sheep and goat LU
	Finishing farms		Others > 8 LU
		Specialised finishers	> 50 males or females > 1-2 year old
		Other finishers	< 50 males or females > 1-2 year old
		Finishing beef and sheep	> 20% sheep and goat LU

LU: Livestock Unit

BFS according to a European livestock zoning

Within the FADN database, each beef farming system can be located in any European country. But it can be useful to place each BFS in a specific zone based on the agronomic potential of the various areas. A. Pflimlin and C. Perrot proposed a division of the EU territory into 7 livestock regions (Figure 1). Those are differentiated according to bio-geographical characteristics (boreal, Mediterranean and alpine), land use (grasslands and part of permanent pasture, maize, crops, etc) and classification of the area as mountain, less favoured or otherwise. This zoning differentiates areas which are more or less specialized in livestock production:

- 3 zones mainly classified as less favoured areas: mountains (including the Centre of France), Mediterranean, and Grasslands regions: 60% of livestock farms and 77% of permanent grasslands;
- a transition area with maize silage and grasslands;

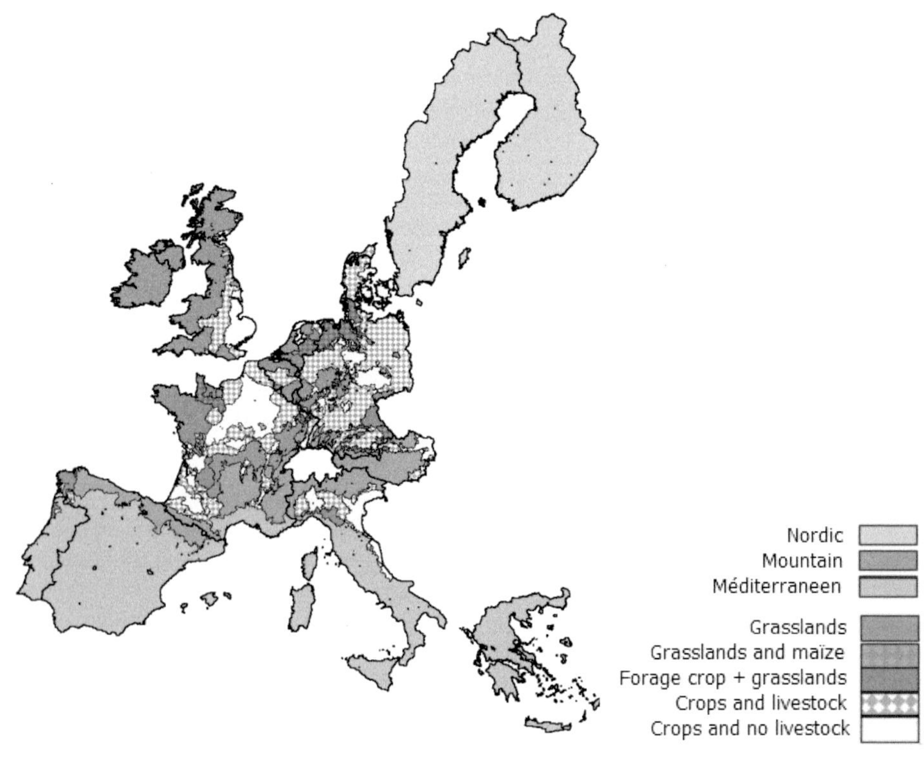

Figure 1. Zoning of EU livestock systems (source: Eurostat - Structure Census 2000 adapted by the French Livestock Institute).

- 2 zones with forage crops and crops and livestock: 25% of the livestock farms, 17% of the permanent grass, and 62% of the maize silage;
- the Nordic specific area.

The diversity of EU BFS - results

For 2004, the most recent year for which FADN data are available, the number of 'bovine' enterprises (holding at least 5 bovines and including dairy farms), was about 921,000 in EU-25.
The level of engagement of those farms in beef production is relatively low as the proportion of total returns from beef ranges from 14% to a maximum of 54%. Within those 921,000 commercial farms, 491,000 farms (53%) are mainly engaged in beef production. Half of these are cow-calf farmers, over one quarter (27%) are specialized in fattening and 20% have coupled dairy and beef production. This classification (Table 2) illustrates the diversity of the systems and their impact on land use and production. The location of the main beef farming systems across the various EU countries is shown in Table 3 and their zoning location is shown in Table 4. The contribution of the different types of systems has been calculated on the basis of the acreage used (total and grass), the total livestock units and the total beef output (in value (€) and net of purchasing).

Table 2. EU BFS characteristics.

Production	Cow calf (CC)			Finishing (F.H)			Dairy		Small farms	Total
BFS (Beef Farming Systems	CC+sheep	CC+Fin.	Pure CC	Fin.+ sheep	Spec. Finish.	small fin.	Dairy+ Beef	Pure dairy		
FADN farms	1,305	1,358	3,182	194	350	1,016	3,420	9,706	1,077	21,608
Enterprises (1000 farms)	49.9	65.0	171.4	9.2	12.7	58.8	123.8	338.8	91.4	920.9
Acreage (ha)	151.9	73.3	69.7	87.9	94.0	45.9	81.5	56.9	19.8	63
% grass on total acreage	88%	63%	69%	68%	45%	42%	62%	61%	33%	62%
% beef on total returns	34%	49%	53%	22%	51%	34%	21%	9%	14%	19%
Labour unit	1.7	1.5	1.4	1.6	2.0	1.4	2.0	1.9	1.4	1.7
Livestock unit	52.5	61.4	47.4	39.5	127.2	31.6	93.6	66.7	3.8	58.1
% BFS farms	5%	7%	19%	1%	1%	6%	14%	37%	10%	100%
% land (on land for BFS)	13%	8%	20%	1%	2%	4%	17%	33%	2%	100%
% BFS livestock unit	5%	8%	15%	1%	3%	3%	22%	42%	1%	100%
% Beef Production	6%	11%	24%	1%	6%	7%	20%	24%	1%	100%
Stocking rate (LU/ha)										
<1.4 LU/ha	65%	50%	72%	29%	15%	56%	31%	34%	60%	47%
>1.8 LU/ha	12%	23%	13%	23%	57%	25%	34%	37%	27%	28%
%OTEX 41-42-43-44-71	85%	63%	66%	57%	53%	45%	84%	86%	25%	71%

Source: EU-FADN – DG AGRI G3

Table 3. Location of the EU BFS per country.

BFS (Beef Farming Systems)	Cow calf (C.C)			Finishing (F.H.)			Dairy		Small farms	Total
Countries	CC+ Sheep	CC+ Fin.	Pure CC	Fin.+ sheep	Spec. Finish.	Small fin.	Dairy+ Beef	Pure dairy		
Enterprises (farms)	49,889	65,100	171,388	9,202	12,656	58,763	123,788	338,725	91,404	
Belgium	-	4%	3%	-	-	-	5%	3%	-	24,375
Denmark	-	-	2%	-	-	3%	-	2%	2%	16,509
Germany	-	9%	4%	-	31%	21%	21%	21%	4%	131,398
Greece	-	-	-	-	-	-	-	-	3%	11,646
Spain	8%	-	19%	-	-	8%	2%	8%	4%	76,182
France	16%	20%	33%	-	4%	4%	30%	20%	4%	192,326
Ireland	27%	27%	15%	30%	-	20%	12%	3%	4%	103,122
Italy	11%	12%	8%	-	21%	14%	7%	12%	34%	116,931
Luxembourg	-	-	-	-	-	-	-	-	-	1,400
Netherlands	-	-	-	-	-	2%	-	7%	-	31,567
Austria	-	-	2%	-	-	7%	9%	8%	2%	51,565
Portugal	-	-	5%	-	-	4%	-	3%	34%	55,716
Finland	-	-	-	-	-	2%	-	4%	-	21,310
Sweden	-	3%	2%	-	-	2%	2%	2%	-	17,957
United Kingdom	33%	15%	4%	48%	22%	6%	5%	5%	-	68,910
Total	100%	100%	100%	100%	100%	100%	100%	100%	100%	920,916

Source EU-FADN – DG AGRI G3
(-: under 2%)

Table 4. Location of the EU BFS per livestock zones.

BFS (Beef Farming Systems)	Cow calf			Finishing			Dairy		Small	Total
EU livestock zones	CC+ sheep	CC+ Fin.	Pure CC	Fin.+ sheep	Spec. Finish.	Small fin.	Dairy+ Beef	Pure dairy	farms	
Crops and no livestock	-	5%	5%	-	11%	9%	3%	5%	10%	5%
Forage crop-grassland	2%	10%	11%	-	-	5%	16%	16%	4%	12%
Crops and livestock	7%	21%	16%	15%	41%	29%	28%	21%	12%	20%
Grassland	60%	43%	27%	73%	28%	27%	25%	14%	6%	27%
Grasslands and maize	-	4%	2%	-	-	5%	7%	10%	-	6%
Mediterranean	22%	8%	20%	-	-	15%	5%	8%	57%	16%
Mountain	8%	5%	16%	-	-	5%	12%	19%	7%	13%
Nordic	-	5%	2%	-	6%	5%	3%	7%	2%	4%
Total	100%	100%	100%	100%	100%	100%	100%	100%	100%	100%

Source: EU-FADN – DG AGRI G3
(-: under 2%)

Small livestock farms

These are numerous (91,000 breeders, 10% of the total) but they control only 1% of the total livestock units and their role in beef production is negligible. In fact, those systems are more involved in other activities and operated by part-time farmers: 75% are classified in other types of farming than livestock. Most of them (57%) are located in the South of Europe and especially in the Mediterranean areas of Italy, Spain or Portugal. There, they contribute to the rural economy and land use.

Cow-calf farms

The bovine activity of cow-calf farms is based on calf production from a suckler cow herd. They represent one third of the bovine farms and control the majority of the EU suckler cows. Actually, their contribution is effective in the EU beef production as well as in the land use: they utilise 41% of the land used by the total bovine herd with a dominance of grass and they are engaged in 41% of the production.

- Most of them are pure cow-calf producers including 170,000 breeders and 19% of the European bovine owners. On grasslands, they produce weaners or backgrounders sold to fatteners. Their herds are rather small with 47 livestock units and less than 35 suckler cows. These farms are mainly specialized in beef production with 69% of their acreage in grass, and they manage the farm at a low stocking rate. Those cow-calf farms are located in 3 areas:
  - in the grasslands of Britain, Ireland, France, and North of Europe (27%);
  - in the Mediterranean areas of Italy, Spain, Greece or Portugal (20%);
  - in the mountain areas of France, Spain and Eastern Europe (16%).
- 65,000 cow-calf producers fatten the majority of the progeny as suckler calves, bulls, heifers or steers on their farms. This fattening enterprise is designed to add value to the weanling production enterprise. With a rather bigger size than the pure cow-calf beef farms in herd and acreage, they manage land more intensively and produce forage crops: 23% operate at a stocking rate higher than 1.8 LU/ha. This system represents 7% of the farms and 8% of the European bovine herd. Their contribution in the fattening phase is important, especially for bull and steer production. They are mainly located in the grasslands of Ireland and U.K. and in the forage crops areas of France and Northern Europe.
- 49,900 farms are cow-calf and sheep producers. Their size (151 ha on average) and stocking rate (65% under 1.4 LU/ha) indicate a relatively low intensity of management. Those systems are specialized in livestock production with 88% of the acreage in grassland. The combination of beef and sheep production leads to good productivity of land use. Those systems are located on big farms producing steers on grasslands in Ireland and in the United Kingdom.

Finishers

Relatively few in number, the farms specialised in the finishing activity are involved in 14% of the EU beef production.
This classification differentiates 3 types:

- 12,700 farms specialised in bull fattening. On an average of 94 ha, they cultivate more crops than forage and have more labour units than the cow-calf producers. In those farms, the fattening enterprise is large-scale (127 LU in average). They control 36% of the livestock units and play a significant role in the beef production of Italy, Germany, UK and Spain.
- 58,800 farms have medium-sized herds of fatteners (less than 50 young males) and have no calf production activity. A large number are rather small fatteners of young bulls with maize silage and concentrates (in the grasslands and crops and livestock areas of Italy, Germany, Austria

and Sweden). In Ireland and UK, those herds are the main European producers of steers on permanent pastures.
- 9,200 beef and sheep farms are involved in fattening with more than 50 young males produced. Those farms are mainly located in grassland areas of the UK and Ireland.

Dairy and beef farms

123,800 enterprises are involved simultaneously in dairy and beef production. These systems represent one quarter of the beef producers and use 17% of the beef farming land. In most of the cases, beef production has been developed on farms with small milk quotas. According to land and labour availabilities, different types of production can be found, like bull fattening of dairy calves on the French or German farms, or like steer and heifer production on pastures in Great Britain and Ireland. Their weight in the EU beef production (20%) shows how important the future of their involvement in production is.

Conclusion

This classification of EU beef systems, differentiating dairy and beef farms and also cow-calf production and finishing enterprises, is instructive and needs further expansion. It gives a good overview of the different beef farming systems found in the European community and highlights the involvement of suckler cow owners and dairy producers in the beef activity. It also shows the great diversity of system types due to the fact that beef production is often carried out in conjunction to other activities such as dairy, sheep or crop production. This diversity is the result of the adaptation of livestock management to forage availability: beef production is successfully managed on grasslands, mountains but also on crops and livestock areas. For instance, the British systems show that steer and heifer production is the best method of exploiting their grasslands. In the South of Europe, another deal is organised between suckler cow owners located on French and Spanish grassland areas and specialised in rearing, and finishers located in crop areas and engaged on a professional basis in fattening of bulls or heifers in Spain and Italy. These complementary productions need to be evaluated through their impact on land use as well as their competitiveness on the market.

References

Chatellier, V., H. Guyomard and K. Lebris, 2005. The diversity of the professional beef farms in UE. INRA.

Pflimlin, A., B. Buckzinski and C. Perrot, 2005. A zoning proposal to characterize the diversity of livestock farming systems and regions in Europe – French Livestock Institute.

Impact of the new CAP on French beef farming systems

P. Sarzeaud[1], F. Bécherel[2] and C. Perrot[3]

[1]*Institut de l'Elevage - Rond point Le Lannou - 35042 Rennes cedex*
[2]*Institut de l'Elevage, Bd des Arcades, 87 060 Limoges Cedex 2*
[3]*Institut de l'Elevage, 149 Rue de Bercy 75595 Paris cedex 12*

Abstract

With one fifth of the European cattle herd held in France, the country appears to be affected by the implementation of the so-called Mid-Term CAP review. The main part of its beef cattle, located in the grasslands of central France, is engaged in calving activity. This double role in production – providing weaners to south European fatteners – and land use –maintenance of mountains and low potential lands – could have been strongly affected by the new principles of premium decoupling and 'subsidiarity'. The French choice to keep the suckler cow premium coupled seemed to help the preservation of the calving potential as far as beef breeding is concerned. The risk is to weaken the lowly integrated fattening activity.

Keywords: French beef production, beef farming systems, policy analysis, farm strategy analysis

The French beef production

With 19 millions of cattle in 2006, the French bovine herd is divided equally between dairy and beef production (around 3.8 million dairy cows and 4 million suckler cows) (see Figure 1). But as dairy cows are mostly located in the west and north intensive areas, the spatial distribution of beef cattle depends on pasture availability for cows and sustainability of crops, such as maize silage, for fattening. Therefore, the major part of the French suckler cow herd is located in the grasslands and mountainous areas of the centre of France. Those farms are mainly dedicated to cow-calf production for exportation to Italy and Spain. A third of suckler cows is situated in the intensive plains of the West and in the grasslands of the North-East. These are more dedicated to bull and fattened heifer production.

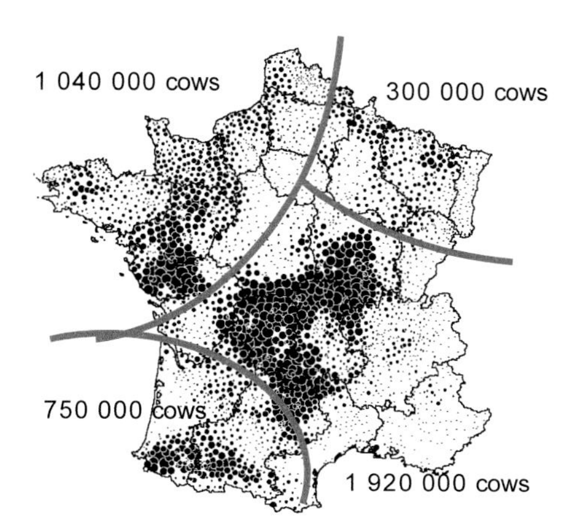

Figure 1. Location of the suckler herd (Census 2000) Data 2005(SCEES).

The French beef farming systems

The beef calving activity is predominant in France: 72% of the beef farms are cow-calf producers and provide the market with weaners or backgrounders. The fattening enterprise as a specialized activity is relatively rare (less than 2000 farms); it is more frequent as a complementary activity to calving or even to dairy activity. Lastly, the local market is supplied by steers and suckler calves. Through a national observatory of bovine systems, managed by the French Livestock Institute and the Chamber of Agriculture, the French beef network provides a good picture of the different types of farms and their characteristics, although their size and economic results are higher than those of the common farms (see Tables 1 and 2). Systems seem to differ in size but also in management. For the main group of cow-calf producers, sustainable herds comprise more than 60 suckler cows per family and use 85% of grasslands. Cattle and land management is relatively extensive. The systems with a mix of activities – calving and fattening for bulls and heifers – are bigger and more intensive, according to crops availability. But the most intensive management is observed in the bull fattening farms where young cattle are fed with maize silage, concentrates or crop wastes. Suckler calf systems are found in the Limousin and south-west areas and are specific to small farms with available manpower; steer producers are very big farms situated in grassland areas of the north-western and north-eastern parts of France (see Figure 2).

Table 1. Different farm types in RA 2000.

Beef farming systems	Number of farms	%
Suckler calf farms	5600	5%
Cow-calf (weaner) farms	52600	51%
Cow-calf (backgrounders) farms	21800	21%
Cow-calf and young bull farms	15600	15%
Steer producers	6100	6%
Bull fatteners	2000	2%
Total	103700	

Table 2. Results of the French Beef Network in 2004.

Beef farming systems	Number	Labour Unit	Suckler cows (bulls)	Cattle Live Unit	Total land use	Forage land use	Stocking rate
Suckler calf farms	24	2	59	75	85	64	1.14
Cow-calf (weaner) farms	84	1.8	64	93	100	79	1.29
Cow-calf (backgrounders) farms	161	1.8	74	119	133	106	1.11
Cow-calf and young bull farms	118	2.0	74	128	131	90	1.54
Steer producers	28	1.8	63	137	145	114	1.25
Bull fatteners	25	1.6	118	68	85		
Total	440	1.8	66	113	122	90	1.45

French Livestosk Institute - Beef Network 2004.

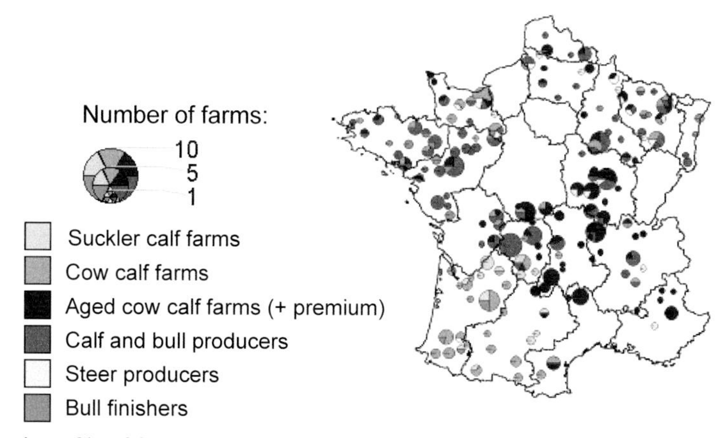

Figure 2. Location of beef farming systems – French beef network.

The main evolution: size increase and specialization of beef production

During the last 20 years, the increase of farm size has been the most important evolution for beef systems: + 2.3% per year between 2000 and 2005 (SCEES). Land availability is linked to the retirement of farmers who are not replaced. An estimation of the French Livestock Institute was that 30% of beef farmers are supposed to stop before 2012, abandoning 15 to 20% of the national herd. The number of dairy farms and suckler cow farms should be below 90.000 in 2015 (see Figure 3). With a relatively low land cost, farms keep on growing and consequently herd sizes are also increasing, although at a lower rate, while stocking rates decrease (-0.08 LU/ha between 2000 and 2003 in the French beef network – French Livestock Institute).

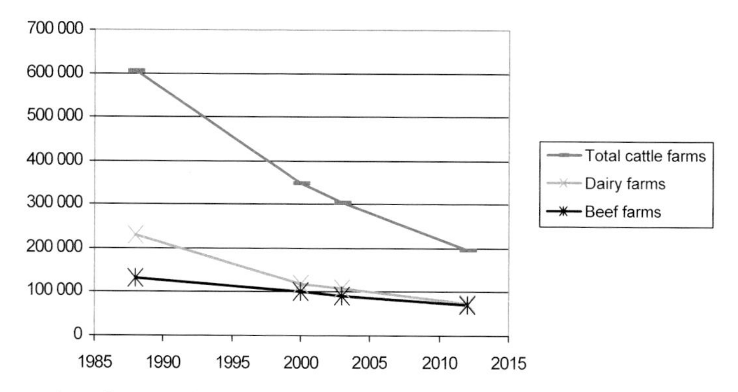

Figure 3. Perspectives for French cattle farms – French Livestock Institute.

The CAP implementation in France

In France, the new CAP became effective in January 2006 with different applications (see Figure 4). The single farm payment (SFP) scheme, with aid no longer linked to production (decoupling), was established for all the farms. SFP levels vary among the farms and depend on the production

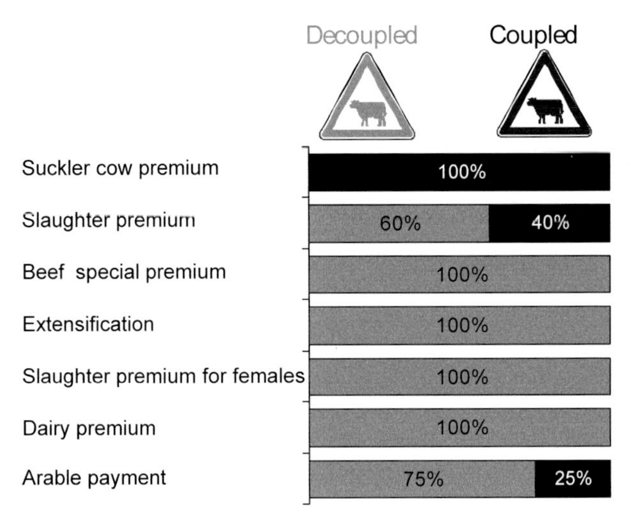

Figure 4. Decoupling implementation in France.

levels during the 2000-2002 period. SFP includes the new dairy premium and the main part of the previous CAP premiums. The high variability of the single farm payments is due to the fact that France did not settle an uniform payment per eligible hectare of agricultural land.
The extensification payment, currently ranging from 40€ to 100€ per animal according to the density, is fully decoupled. Taking into account the role of beef production in territory maintenance, France decided to leave the whole suckler cow premium and 40% of the slaughter premium coupled. And in order to prevent some areas from becoming completely fallow, 25% of the previous arable payment remains coupled to production.
The single payment is moderated by a modulation of 3% in 2005, 4% in 2006 and 5% from 2007 to 2012 for all the farms except the smallest ones. To receive subsidies, farmers must be attentive to recommendations about animal welfare and environmental conditions. One of the main cross-compliance requirements concerns permanent pasture maintenance. Since 1999, modulation has been introduced to support regional projects through the implementation of land management contracts (contrat territorial d'exploitation). Cattle farmers in the mountain regions and pasture areas could benefit from this program that ended in 2006.
Fruit of the previous CAP, subsidies contribute largely to the total return of the beef farms (from 25 to 35%) and are predominant in the income constitution. The CAP reform showed differences in the premium amount between producers. The premium amount per ha depends on the production and the intensification level: for example, from 250-300 €/ha for cow-calf producers to 500-550 €/ha for fatteners (see Figure 5).
Nevertheless, due to the French choices, beef farming systems have not been concerned in the same way by the decoupling act. Cow-calf producers keep a main part of the support received linked to the number of cow premiums and consequently their income is relevant to their production. On the contrary, the bull finishers' activity is mainly decoupled and fatteners could decide more easily to stop or increase their activity without losing outputs.

First intentions

In 2005, a national poll was organized to investigate the new objectives of beef farmers within the new CAP context (see Figure 6). Eight scenarios came up:

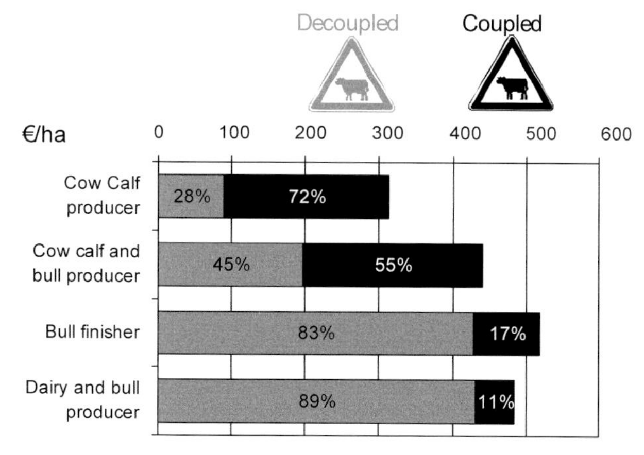

Figure 5. Single farm payment (in red) and coupled premium (in blue) per ha for 4 cases of beef farming systems.

1. Increase the size of the farm and/or the cattle herd.
2. Move towards an intensification of the system by increasing yield per ha or per labour unit.
3. Move towards an extensification of management.
4. Maintain the same management of the farm.
5. Change beef production (e.g. from fattening to cow-calf production).
6. Improve income by developing beef instead of dairy production.
7. Develop new agricultural activities (crops, tourism, etc).
8. Stop beef production.

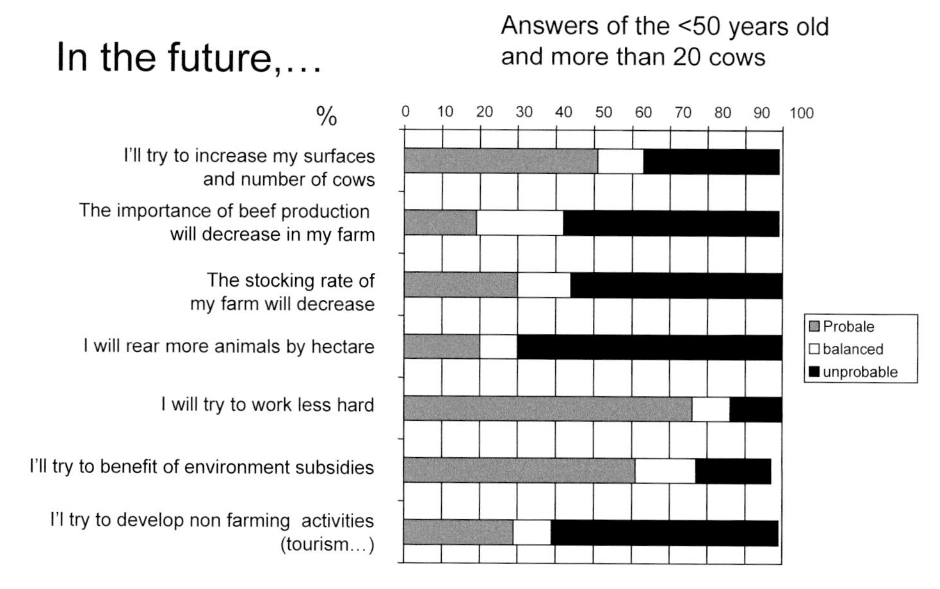

Figure 6. Results of the poll for the beef 'hard core' panel in 2005.

Some of these objectives appeared really predominant for beef farmers; improving working conditions particularly dominates in their wishes for the future. But, as far as the beef activity is concerned, the growth of the herd is also expected to maintain returns. If in the past most of the farmers were interested in moving to extensification management, the end of limit for stocking rate leaves some freedom to return to livestock intensification.

Impact of the CAP on economic results and on production location

To evaluate the impact of the new CAP on the beef farming systems and their management, some simulations have been carried out by the French Livestock Institute on a panel of typical cases: cow-calf producers, cow-calf and bull producers, bull fatteners with no change in the management or with some evolution such as increasing land and herd size, decreasing or increasing the stocking rate, changes in the production: give up or increase the bull fattening activity. According to the situation of the European beef market, in particular a structural deficit prospect in the following years, the baseline chosen kept the hypothesis of maintaining the prices at the same level as those observed in the last years (2003-2004).
So the main impact for the income of the beef systems is relative to the modulation of subsidies at the farm level: 5% of income lost for 5% of modulation. And in the same market context, there are no other consequences for returns or incomes.
Through the analysis, the decision to maintain coupled the suckler premiums appears as a support to the cow calf production. Those farmers are encouraged to keep their suckler cows or even to increase their herd to recover their income. This act helps the mountain and grassland areas, where most of the national herd is located, to keep their suckler cows and not change to an extensive livestock or forest enterprise. In those regions, fattening heifers and young bulls or producing steers is less interesting than before. On the contrary, the decoupling (total or in majority) of premiums related to fattening enterprises, such as the slaughter premium or the bull special premium, seems to give more freedom for activity. The economic interest of fattening appears more reliable on the market situation.

Consequences for each system

Consequences for cow-calf producers

For cow-calf producers, the main evolution concerns the size of the farms and the type of weaners to produce (age of sale, management, weight, and period). As far as premiums are concerned, the decoupling of the bull special premium is supposed to help farmers choose between two production options:

- Heavy and old weaners (more than 450 kg live weight) produced on pastures in the Charolais basin for Italian fatteners. These weaners had obtained the premium previously and can be fattened under quick management. The Italian market demand is oriented to lighter weaners (420 kg) in order to improve the homogeneity of the fattening (feeding, sanitary rules), but there is also a market for non fattened young bulls (rustic or Charolais breeds) and heavy weaners 'pre-fattened' in France.
- Young weaners from 5 to 6 months (250-280 kg) before weaning for Blonde d'Aquitaine or 8 months (300 to 350 kg) for Limousine at the age of weaning for the Italian market. These suit also French fatteners' demand.

Some of the current problems of Italian fatteners are: i) to ensure the supply of weaners all year round; ii) to decrease the price of weaners: and iii) to receive guarantees on the way the calves are produced in France, according to the feed (non GMO food) and the sanitary treatments. The first constraint could force suckler cow producers to change the time of calving (earlier or later in the Charolais and rustic basins) or to maintain a certain variability of weaners.

Consequences for fatteners

The typical French system mixing cow-calf activity and fattening is particularly concerned by decoupling. The recent market evolution for cow-calf, with the price rising due to a good demand by Italian and Spanish fatteners, makes farmers reconsider fattening their weaners. Other considerations are also included in the debate, such as labour cost or specialisation of the farm in case of increasing the size. French fatteners are not numerous and they are relatively small. Only 5% of them keep more than 80 cattle and contribute to 21% of the production. Their difficulty is to improve the fattening interest with relatively low sale prices and to reach a size big enough to reduce their production costs.
In both situations, the point is more or less the opportunity cost of the enterprise compared to arable crops when feeding is based on maize silage and concentrates produced on the farm. In Figure 7, some equivalences are given between purchase price (weaner price) and sale price for a given fattening management and with relatively common efficiency to obtain a crop gross margin. For two years, price balances have been in favour of fattening for beef and also dairy breeds. However, this approach is a limited and specific analysis. Other elements must be integrated in the debate, such as work and any other costs related to the development project.

Conclusion

It is difficult to make any statement on the consequences of the new CAP on beef farming systems, especially because of its recent implementation. On the other hand, part of the evolution is due to other factors such as labour productivity improvement, replacement of farmers, shift of certain types of production such as dairy to beef production, etc. A study led by the French Livestock Institute estimates the potential of growth of beef farms to be constant in the future (2003-2012) compared to the past (1988-1997) (see Figure 8).

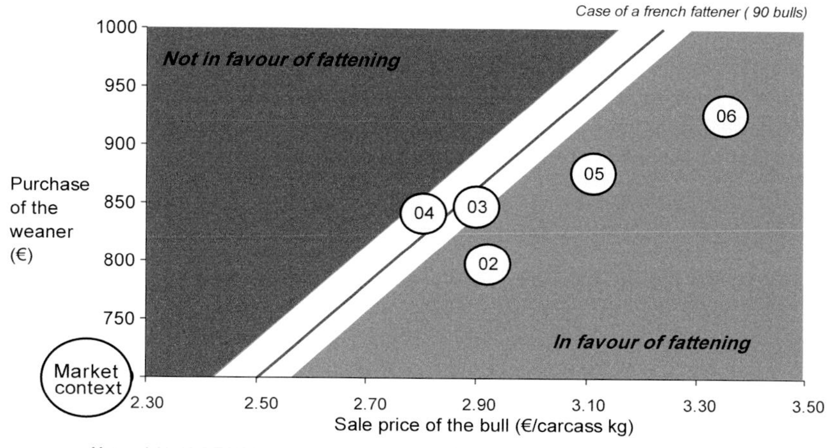

Figure 7. Grid of equivalence between purchase price and sale price to get a gross margin equal to crops activity.

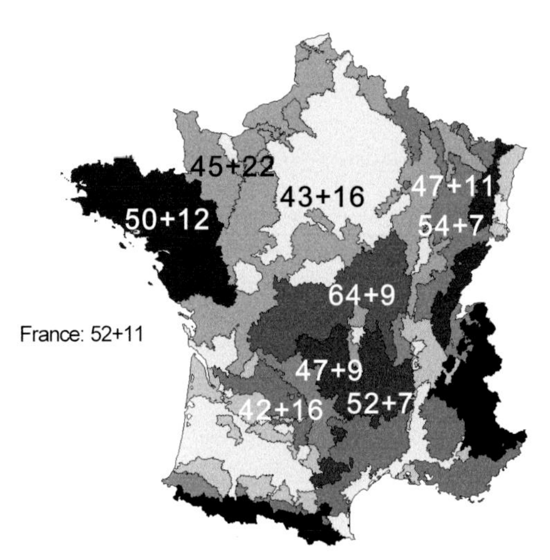

Figure 8. Growth perspectives for each farm remaining between 2003 and 2012 in case of total replacement: SIZE IN COWS 2003+GROWTH (Beef farms with more than 20 cows).
From 2000-2003 'structure' survey French Livestock Institute.

What is certain is that, with the decoupling of the premiums, the new CAP forces the farmers to comply with the market. And fortunately, the market conditions for beef improved much in 2005 and 2006. For example, the price of 'R' bulls increased by 18% compared to 2004, while the price of Charolais weaners increased by 17%. The same stands for the female market.
This context has motivated some farmers, especially in plain areas, who wished to develop fattening production without premiums. This trend is encouraged by the industry, which is worried about reducing its activity and sees bull meat assume a good place on the national market. Some projects, aimed at doubling bull or heifer production, need to undertake new housing investments. In such a context, decoupling is not the issue: the single payment is used as a contribution to the investment capacity. But this situation is relatively limited and shall not be sufficient to boost French bull production.
If the shortage of European production increases, beef from overseas countries shall easily enter the competition, benefiting from its low production cost. Actually, the new CAP can accelerate the specialisation of beef farming systems in relation to the production areas: cow-calf production in the centre of France, linked with the fatteners of southern Europe, and fattening in the plains, as much as possible comparatively to the opportunity of crop production including energy purposes.

References

Brouard, Sylvie and the Limousine Beef Farm Network from the French Livestock Institute and the Chambers of Agriculture, Communication 15th September 2005, Adaptation's strategies of the farms from Limousine Basin.

French Livestock Institute, 2002. The beef, sheep and goat livestock activity from the 2000 census. Herds, farms and production. In 'Economie de l'Elevage' N° 318, 68 p.

French Livestock Institute, 2006. The French livestock activity to the 2012 objective: farm adaptation to the M.T.R. CAP reform? In 'Economie de l'Elevage' N° 353, 82 p.

French Livestock Institute, 2007. Beef farming systems in France: technical and economical benchmarks. Analysis of the Beef Farms Networks results in 2005. 24 p.

Sarzeaud, P., and the West Beef Farm Networks from the French livestock Institute and the Chambers of Agriculture, 2005. CAP reform for the beef farming systems: issues and farm managing adjustments. 36 p.

The EU CAP-reform of 2003 and its consequences for German beef farmers

C. Deblitz[1], M. Keller[2] and D. Brüggemann[3]

[1]Asian Agribusiness Research Centre, Charles Sturt University, Orange, NSW 2800, Australia; claus.deblitz@fal.de
[2]Auf dem Berg 15, 46514 Schermbeck; M.A.Keller@gmx.net
[3]Institute of Farm Economics, Federal Agricultural Research Centre (FAL), Bundesallee 50, 38116 Braunschweig, Germany; daniel.brueggemann@fal.de

Abstract

Beef production in Europe is among the sectors most affected by the recent CAP-reform. German beef production is mainly based on intensive bull finishing with origin from the dairy herd. The implementation of the CAP-reform varies widely between the member states and the German government opted for a full decoupling of payments and a 'dynamic hybrid payment model'. The historical, individual farm-based Singe Farm Payments (SFP) are gradually converted into homogeneous, regionalised acreage payments for both cropland and grassland in the period 2010 to 2013. The impact of the CAP-reform is shown for two example farms, both sourced from the typical farm sample of the agri benchmark project. The farm income of the intensive beef finisher in West Germany decreases dramatically over the adjustment period whereas the farm income of the extensive suckler-cow farm in East Germany increases slightly when compared with the Baseline scenario representing the pre-reform Agenda 2000 policy framework. The main reason is that the increasing acreage payments for the cropland do not compensate for the 'loss' of the individual SFP of the intensive finishers whereas the newly introduced grassland premium for the grassland farm compensates the 'loss' of the individual SFP. Rising rent prices will, however, reduce the future farm income. Based on these findings, farm strategies for a number of intensive finishers were identified, specified and analysed in cooperation with farmers and their advisors. From an income point of view, the stop farming strategy was unfavourable unless surplus family labour can be used in the local labour market. For most of the farms, the moderate growth scenario proved more favourable and superior to the strong growth scenarios. This conclusion was supported when including risk in the analysis. The moderate growth scenario showed among the highest income expectations and the lowest probabilities for making a loss.

Keywords: beef production, EU CAP-reform, policy analysis, farm strategy analysis, risk analysis

Introduction

Situation and objectives

With tariff rates of approximately 90 percent and a sophisticated set of direct payments, beef production in the European Union has been one of the most protected and subsidised agricultural sectors. It is therefore not surprising that the beef finishing and cow-calf sector is among the most affected by the 2003 CAP-reform, also known as Mid-Term Review (MTR). This holds particularly true for intensive bull finishers with high stocking rates and limited land availability.
The goals of this paper are to:

- examine how the MTR was implemented in Germany;
- analyse the consequences of the reform for different farm types in Germany;
- identify and analyse adjustment strategies for intensive bull finishers;
- to evaluate the risk involved in the implementation of the adjustment strategies.

Methods and working steps

Data and information about the characteristics and the implementation of the MTR were obtained from literature review and the European Commission (2005). All farm level and production system information is derived from the agri benchmark project and related master and diploma theses (Keller, 2006 and Brömmer, 2005). agri benchmark is a world-wide project of farm economists with participation of farmers and advisors. The main objective is to generate sustainable, comparable, quantified information about farming systems, their economics, their framework conditions and perspectives world-wide.

For quantitative farm level analysis, typical farm data from the agri benchmark farm sample as well as individual farm data from beef finishers are used. The analysis was performed in close cooperation with producers and advisors, using the farm level simulation model TIPI-CAL. A detailed description of the typical farm approach is provided by Deblitz and Zimmer (2005) as well as on the agri benchmark website at www.agribenchmark.org. For reflecting risk in the farm strategy analysis, the Excel add-in SIMETAR software developed by James Richardson (Texas A&M University) was used.

The farm level data and results obtained are not representative but reflect a fair proportion of the beef finishing and cow-calf production systems under operation. To allow conclusions about other farm types, general considerations about the impact of the policy and different strategies on different types of farms are provided where relevant.

The following working steps are undertaken: The first chapter provides an overview of the German beef production and its situation in Europe. In the next chapters, the implementation of the MTR in Germany is described and reasons for and general consequences of the different ways of implementation are presented. The following chapter is dedicated to the consequences of the MTR on beef finishing and cow-calf farms in Germany, based on quantitative analysis of two typical farms taken from the agri benchmark sample as well as general considerations for further farm types. Based on the results of this chapter, a farm strategy analysis for intensive bull finishing farms is performed in the penultimate chapter, which is further detailed by reflecting production and market risk in the last chapter.

Status quo of beef production in Germany

Germany's beef production position within Europe can be summarised as follows:

- With a production of approximately 1.3 million tons, Germany is the second largest beef producer in the EU-25, which has a total production of approximately 8 million tons (ZMP, 2006).
- With a per capita consumption of approximately 12.5 kg per year, Germany has the lowest beef consumption in the former EU-15 (ZMP, 2006). Consumption is only lower in the New Member States.
- Germany is a 'milk' country, i.e., 88 percent of the total cow numbers are dairy cows (ZMP, 2006). Thus, the vast majority of the beef produced in Germany is from dairy origin. Main breeds include Holstein (mainly in the north of Germany) and Simmental (mainly in the south of Germany).
- The prevailing production system is bull finishing on a corn (grass) silage plus grains/concentrates/soybean ration in confined barn systems (Brömmer, 2005).

- Productivity levels are rather high with daily weight gains (DWG) of around 1,000 g per day for Holstein bulls and 1,100-1,300 g for Simmental bulls. Final live weights are at 620 kg for Holstein and more than 700 kg for Simmental bulls (Brömmer, 2005).

The implementation of the CAP reform in Germany

Germany opted for a full decoupling of the direct payments and dynamic hybrid (combi) model as follows:

- For each farmer, a part of the decoupled payments is paid as a SFP based on the historic payments he received and another part is paid on an acreage basis. Figure 1 shows which part of the formerly coupled payments were converted into SFP, arable and (as a newly introduced payment) grassland payments, respectively.
- The acreage payments are homogenous for defined regions (Bundesländer) with slight differences between the regions. Further, in the beginning of the implementation, there are different acreage payments for cropland and for grassland (for regionalised acreage payments see Table 1).

From 2005 to 2009, the relation between the SFP and the acreage payments will remain constant. In the four-year period 2010 to 2013 the following two main changes are going to occur in the system:

1. The SFP will gradually be phased out (to zero) in four steps in favour of the acreage payments.
2. The grassland payments will gradually increase until they reach the level of the cropland payments.

As a consequence, in 2013 there will only be homogeneous acreage payments for cropland and grassland while slight differences between the Bundesländer will remain. The initial cropland premium may increase or decrease in that period depending on a) the share of SFP in the total payment amount and b) the share of grassland in the Bundesland. Table 1 shows the development of the acreage payments per Bundesland and Figure 2 illustrates the evolution of acreage payments for cropland and grassland, taking Bavaria as an example.

The background and intention of policy when designing the German hybrid payment model had three key considerations:

1. **Avoid court cases between neighbours**. In the historic payment situation, the hectare equivalents of the payments received used to be largely heterogeneous between neighbouring crop farms

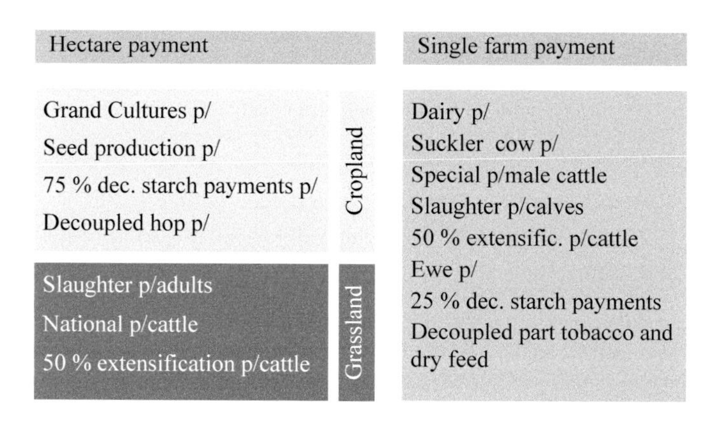

Figure 1. Conversion of coupled payments into decoupled single farm payments and acreage payments (DBV, 2003).

Table 1. Start and end values of acreage payments by regions (Bundesländer) in Germany (BMVEL, 2004).

Bundesland	2005		2013	2013 vs. 2005
	Grassland	Cropland	Homogeneous acreage payment	Direction crop payments
Baden-Wuerttemberg	56	317	302	-
Bavaria	89	299	340	++
Brandenburg	70	274	293	++
Hesse	47	327	302	--
Mecklenburg-Western Pomerania	61	316	322	+
Lower Saxony	102	259	326	+++
North Rhine-Westphalia	111	283	347	+++
Rhineland-Palatinate	50	288	280	-
Saarland	57	296	265	---
Saxony	67	321	349	++
Saxony-Anhalt	53	337	341	+
Schleswig-Holstein	85	324	360	++
Thuringia	61	338	345	+
Germany	79	301	328	++

Changes in crop payments relative to their initial values:
– – – much lower, – – moderately lower, – slightly lower, + slightly higher, ++ moderately higher, +++ much higher

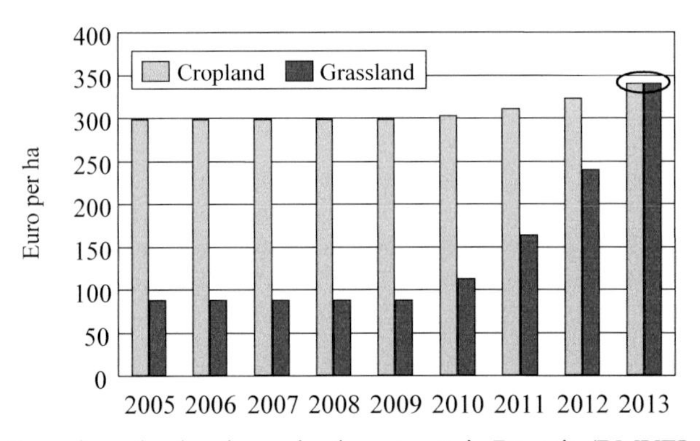

Figure 2. Evolution of cropland and grassland payments in Bavaria (BMVEL, 2004).

and intensive livestock farms with the same natural conditions. Payment levels were usually around €350 per ha for the cash crop farms and equivalent to €500-900 for the beef finishing farms when converting the per head payments received to the hectare. With a SFP only, both neighbours would have maintained these levels of payment which many politicians thought inequitable and might lead to legal action if the (possibly former) cash crop farmers would seek the same payment levels as their (possibly former) beef producing neighbours. As a result, it was concluded that a homogeneous acreage payment for each hectare of land would be the more viable option.

2. **Minimise liquidity problems of intensive beef farms**. By introducing a homogeneous acreage payment exclusively from the first day of implementation, intensive beef finishers could have

suffered from severe liquidity problems because their hectare equivalents would have dropped from the originally individual high values to the low values as shown in Table 1. Therefore, policy opted for the hybrid (combi) model and the phasing out of the SFP after a constant relation between the SFP and the acreage payment for five years (2005-2009). This minimises large changes in liquidity and provides intensive livestock farms time to adjust.

3. **Promote grassland**. The then government coalition of social democrats and greens wanted to promote grassland which has never received support from the first pillar measures by introducing a grassland premium. The grassland premium is to increase up to the same level as the cropland premium in the final year of implementation (see Table 1 and Figure 2).

Consequences for beef farms in Germany

With decoupled payments, the future profitability of beef finishing and cow-calf production will solely depend on the price relations rather than subsidies. In case producing beef (or keeping suckler-cows) without direct payments is no longer profitable, one should better stop this form of production.
The profitability of beef farming is additionally influenced by cross-compliance regulations, where land farmed must be kept open to be eligible to receive payments. On flat land this can be done mechanically by mowing/mulching the land once or twice a year at costs of approximately €40 to 60 per hectare for one mowing.
However, stopping farming usually bears some cost as well. Interest and principal for loans must be paid, certain taxes, duties and levies, as well as insurances, must be reflected and maintenance for buildings must be taken into account – if they can not be sold, alternatively used and if legislation requires to maintain them (for example old buildings). Furthermore, the cost of possible degradation of the land and its restoring should also be reflected in these considerations.
The fulfilment of the cross-compliance regulations is however the reference system for all alternative land uses. On a per-hectare basis, the loss per ha on a total cost basis may not be more than the cost for mowing/mulching the land plus the overhead cost mentioned above. Otherwise mulching would be the better alternative.
2005, in fact, when compared with the previous year, recorded a ten percent decline of beef production in Germany with a total of 1,216 million tonnes. In 2006, production rose again slightly to a total of approximately 1,237 million tonnes (plus two percent) (ZMP, 2007). At the same time, beef prices increased significantly and calf prices were on the rise too, however to a lesser extent, so that the profitability (without direct payments) improved. Additionally, individual payment levels for almost all farms remained rather constant around the level of the year 2004 resulting in higher returns when considering the whole farm level. This means that the economic situation in many farms improved in 2005 compared with 2004.

Typical farm case studies

Two significantly different typical farms from the German agri benchmark network were chosen to illustrate the different impact of the CAP reform on different production systems.
Farm 1: A specialised, intensive **bull finisher** in the West of Germany producing 260 bulls per year on a corn silage, concentrate and soybean ration. Apart from the beef, the farm sells wheat and triticale which are not used for feeding.
Farm 2: A **cow-calf producer** in East Germany with a herd of 1,100 suckler-cows producing baby-beef from weaned calves as well as weaner calves for further finishing on other locations. The farm is 100 percent grassland-based and located at the Baltic sea coastal region in North-East Germany.
For both farms, the following analysis was performed:

1. A reference system (Baseline) for comparison with the MTR was defined. The baseline reflects the most likely situation if the reform hadn't taken place and therefore would portrait the continuation of the Agenda 2000 policy. It is characterised by the continuation of the coupled payments, constant production and productivity, as well as constant prices for beef and calves based on 2004 values, as well as increases in general costs such as labour, energy, machines, and general overhead costs. This explains why the farm income decreases over the course of time.
2. Two MTR (CAP-reform) scenarios were calculated both reflecting the decoupling of the payment systems with two different sub-scenarios on prices:
   a. the same price assumptions as in the Baseline, i.e. constant beef and calf prices based on the year 2004. This scenario was calculated to exclusively illustrate the policy-impact albeit it is already passed by price increases reflected in the next sub-scenario:
   b. the same productivity and production assumptions as in the Baseline but reflecting the price increases for beef and calves that took place in 2005, i.e. constant beef and calf prices based on the year 2005.

Figure 3 shows the development of the (whole) farm income of **Farm 1** from 2004 to 2013, the year in which the reform is fully implemented. The farm income includes the decoupled payments and therefore does not show the profitability of the beef finishing enterprise. The chart shows the comparison between the Baseline and the two MTR-scenarios. There are three main milestones for the development of the farm income:

- **The year 2005**. Without the price increases, the farm income would have been slightly below that of 2004. The reason is that the regionalised acreage payments and the SFP could not compensate for the loss of the previously coupled payments (see Figure 1 for conversion of historic payments). In other words, the total payments received by this farm in 2005 were less than in 2004. In the second MTR-scenario, the profit increases because the price increases for beef compensate for both the decrease of the payments and the increase in the calf prices.
- **The year 2010**. This is the year in which the conversion of the SFP into the acreage payments begins. The total payments (and thus the total returns) decrease significantly because the 'loss' of the SFP can not be compensated by the minor increase of the acreage payments.
- **The year 2013**. In this year, the farm income is approximately €20.000 lower than in the Baseline scenario of the same year.

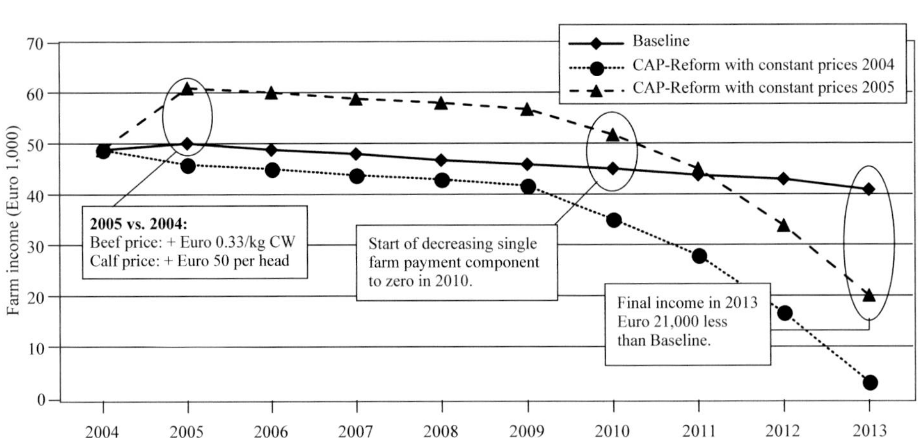

Figure 3. Development of the farm income (without interest on payments) for an intensive bull finisher with 260 bulls annual production (Keller 2006, modified).

Figure 4 shows the projection of the (whole) farm income including the decoupled payments for **Farm 2**. It is important to mention that it does therefore not show the profitability of the cow-calf enterprise. Again, three milestones are identified:

- **The year 2005**. Without the price increases, the farm income would have been significantly below the income of the year 2004. There are two reasons for this, namely:
  1. The 'loss' of historic payments that have been turned into the regional acreage payments (50% of the extensification premiums plus the total of all slaughter premiums, adding up to almost €100,000).
  2. The new grassland premiums are low in the initial phase of implementation. In other words, the total payments received by this farm in 2005 were significantly less than in 2004. In the second MTR-scenario, the profit increases because the price increases for the weaner calves and beef, thus compensating for the decrease of the payments.
- **The year 2010**. This is when the conversion of the SFP into the acreage payments starts. The result of this transformation is significantly different from Farm 1. The reason is that the share of grassland is 100 percent which until the end of the four-year period receives significantly higher premiums every year. As a consequence, the farm income increases in this period.
- **The year 2013**. The farm income in the year 2013 is slightly higher than in the Baseline based on constant prices of the year 2004. In 2013, all grassland of the farm receives an acreage payment of approximately €322 per hectare. With a total (grass)land area of 1,335 hectares this provides a new payment income of almost €500.000, which compensates for the 'loss' of the previously coupled livestock payments. Taking the market price increases in 2005 into account, the farm income is projected to be a further €100,000 higher than in the Baseline.

It should be mentioned that the positive results for the grassland farm can only persist if rent prices remain stable. It can, however, be assumed that the grassland premiums which have been implemented as part of the reform, will lead to further substantial increases of rental prices. As this is a long-term aspect, it is not discussed further here.

The fact that the overall farm income is likely to be equal or even higher than in the Baseline does not necessarily mean that the continuation of cow-calf production is profitable because the decoupled payments may not be considered as returns of the cow-calf enterprise. The same is generally valid for farms of type 1 analysed here.

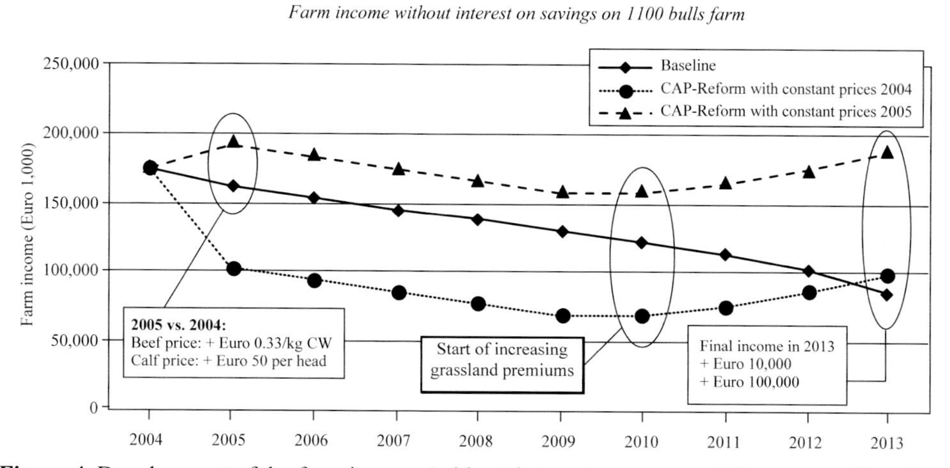

Figure 4. Development of the farm income (without interest on payments) for a cow-calf producer with 1,100 suckler-cows (own calculations).

Further conclusions

Cost of production analysis performed in the agri benchmark project (Deblitz, 2006) suggests that with the decoupling of the payments, the mid- and long-term profitability of beef and cow-calf production is not sustained. The fact that the decrease of production to date has been rather moderate, is possibly due to the following:

- Due to the rise in market prices and the fact that the total income in most farms will not significantly decrease until 2009, it can be assumed that a major drop in production can only be expected from 2010 onwards. This reflects the perception of some farms that if the total income remains high, one will continue with current production, despite decoupled payments.
- Given the presently high price levels, the production of beef and weaner calves in many cases may be profitable when only considering the marginal (or short-term) costs of production. This changes however as soon as investments have to be made or a long-term perspective is taken.
- As mentioned, giving up production has a certain cost, too (see chapter 4 'Consequences for beef farms in Germany' and chapter 5.4 'Results of the strategy analysis', stop strategy).
- Many farmers might wait for others going out of business before they do and thus speculate on higher market prices due to reduced supply.
- Especially for many small cow-calf producers in West Germany, long-term profitability is likely to not be the driving factor for keeping suckler-cows.
- Due to their legal status and social obligations against their employees, a significant number of large farms in East Germany are likely to use a part of the decoupled payments for compensating losses that occur without the decoupled payments.
- Some, especially older farmers, might not have employment alternatives outside their farm. They might continue farming on a marginal cost approach until they retire.

Strategy analysis for intensive bull finishers

The previous chapters illustrated that intensive bull finishers are those who are going to face significant income losses once the SFPs are phased out. In addition, the current expansion of bio-energy plants is going to have a negative impact on the profitability of beef production as it is likely to result in stable or even increasing land prices. Furthermore, an increase of opportunity costs for feeding corn silage to beef cattle can be expected if biogas plants are able to pay above present (forage) market prices. This issue could, however, not be reflected in the following calculations. It is the subject of further research currently conducted within the agri benchmark project.
Based on the findings of the policy analysis performed above, the next logical step was to consider adjustment strategies and ways to avoid the economic decline which is likely to affect intensive beef production in the near future. This was done in close cooperation with farmers and advisors and using the methods and tools applied in the agri benchmark project (Keller, 2006).

Case study farms

Five farms were analysed, all located in the main bull finishing region in the West of Germany, Northrhine-Westphalia. They all run a similar finishing system based on corn silage plus grain / concentrate / soybean ration. To identify possible differences in the policy and farm strategy impacts, different farm sizes, breeds and start age of the finishing animals were chosen. The farms can be characterised as follows with the number indicating the total number of bulls sold per year:

- DE-160: Finishing Holstein bulls from calves (14 days old, 55 kg).
- DE-160F: Finishing Simmental bulls from stocker cattle (4 months, 175 kg).

- DE-260: Finishing Simmental bulls from stocker cattle (4 ½ months, 185 kg).[2]
- DE-320: Finishing Simmental bulls from calves (43 days, 90 kg).
- DE-515: Finishing Simmental bulls from stocker cattle (4 months, 166 kg).

Due to the limited scope available in this paper, the farm DE-260 was selected to illustrate the results of the analysis. Whenever significantly different, results for the other farms are mentioned separately but not shown in figures.

Policy analysis

A policy analysis was performed for all farms, basically yielding the same conclusions as for **Farm 1** analysed above, i.e. a dramatic decline of farm income until 2013. Differences among the farms mainly occur depending on the share that the previously historic payments converted into the SFP (i.e., the special premium) had before the reform was implemented. Farms which had a high share of the special premium in the total premiums – mainly those characterised by a relatively low stocking rate compared with their colleagues – are better off at the beginning of the implementation and suffer from bigger premium losses at the end of the transformation of SFP into acreage payments and vice versa.

Defining strategies

In a feedback procedure with the five farmers and their advisor, possible adjustment strategies were defined which were then analysed for all five farms to ensure comparability of the results between them. All farmers agreed to the five strategies listed below. The strategies were then individually specified for each of the farms. It was agreed that all strategies should be implemented in the year 2006, i.e., one year after the implementation of the reform. The underlying price and policy assumptions were identical with the scenario 'CAP-reform with constant prices from 2005' shown in chapter 'Typical farm case studies' (constant prices from 2005 based on a price increase from 2004 to 2005).
The following strategies were identified:

- Strategy 1: **No adjustments**: Produce the same number of animals; no changes in the production system (equivalent to the second MTR-scenario in the policy analysis).
- Strategy 2: **Stop beef farming in the year 2006**: Animals on the farm stay until they are finished, no new calves are bought; no production on the land anymore; fulfilment of cross-compliance regulations by mulching the land once a year; sell all machines not used for mulching; keep all rented land and obtain payments for it; leave buildings empty but maintain them according to legal requirements (particularly relevant for old buildings).
- Strategy 3: **Strong growth**: Increase of stock numbers between 30 (larger farms) and 100 percent (DE-160 and DE-160F): major investment into new barns, financed with loans; conversion of existing cash-crops (wheat, other cereals) into corn-silage; if necessary, renting of additional land for corn-silage at local rent prices; additional field work performed by contractors at regional rates; additional work in the cattle herd performed by family members (if available) plus hired workers at local wage rates; adjustment of overhead costs to the number of animals, if appropriate.
- Strategy 4: **Moderate growth**: Growth with own financial and labour resources, mainly by change in use or extension of existing buildings; additional corn silage is obtained by converting existing cash crop; if after that not enough land is available, land is rented; additional field work is done with the family labour in case of replacing cash crops by corn silage; field work for additional land rented is contracted out; additional work in the cattle herd is mainly performed by family

[2] The results for this farm's policy analysis were already pointed out in chapter 'Typical farm case studies'.

members; hired labour only in exceptions and with a low share; no investment in additional machinery made; adjustment of overhead costs to the number of animals, if appropriate.
- Strategy 5: **Strong growth + improvement of performance**: In this scenario, the growth strategy 3 is complemented by improvements of the daily weight gain (DWG). Maintaining the duration of finishing, the results are higher final weights. This approach was supported to be realistic by the participating advisors and farmers. The additional DWG varies among the farms from zero to 50 g per day depending on their current performance and intensity levels. Higher DWG can be obtained by improved forage quality, ration optimisation by splitting the herd up into homogeneous groups (size effect), improvement of animal comfort (mostly possible in new barns), and a better health status of the calves. To achieve higher DWG, feed costs (mainly for corn silage) were increased accordingly.

Results of the strategy analysis

The strategies were implemented and calculated using the farm-level simulation model TIPI-CAL. In a first run, a deterministic analysis of the strategies described above was performed. The results are shown in Figure 5 for DE-260.
The **first strategy** reflects the MTR scenario from the policy analysis and is used as the reference strategy under MTR-conditions.
The **second strategy** illustrates the consequences of terminating beef farming. In the long-term, it clearly reveals a negative result because termination of farming still has some costs (see chapter 'Consequences for beef farms in Germany') that are to be covered while returns from beef finishing are down to zero. The acreage payments alone can not compensate for the costs inhered in this strategy. In the first year of this scenario, a jump in farm income can be observed which results from the sale of the machinery. The accumulating interests on savings that can result from this machinery sale are, however, not reflected here as it is assumed that the money obtained here is not re-invested in the farm. Further, in this strategy almost the complete labour (family and hired) is set free. The cash effect on farm income is almost zero because a) family labour is not reflected in the farm income and b) the share of hired labour is negligibly low. In farms with mainly hired labour, the stop of farm activities in many cases would provide the best strategy as all the wages can be saved (with all the social implications this might have).
The **third strategy** is the strong growth strategy. In the first year of implementation, there is a strong drop in farm income due to the purchase of the additional calves (with no beef returns compensating them in this year), additional depreciation for the new buildings and interest payments for the loan that is taken. In the following years, the farm income improves due to the additional beef sales but it does not reach the level of the first strategy due to the high costs involved.
Strategy 4 is the moderate growth strategy with a negative, but less profound, impact in the first year of expansion. In the following years, the farm income becomes even higher than in all other strategies. This is particularly due to the farm's specific relatively low investment costs for the barn extension. It should be mentioned that in farms where higher investment costs occur, this strategy is still superior to the strong growth strategy but does not provide higher farm income than strategy 1.
Finally, **strategy 5** represents the strong growth plus an improvement of the daily weight gains which was defined to be 30 g per day to 1,190 g per day, adding between 14 and 16 kg to the final live weight.

Reflecting risk in the analysis

In the next step, stochastic variables were introduced into the calculations to estimate the risk associated with the different strategies and to get a better idea about the probability of losses and

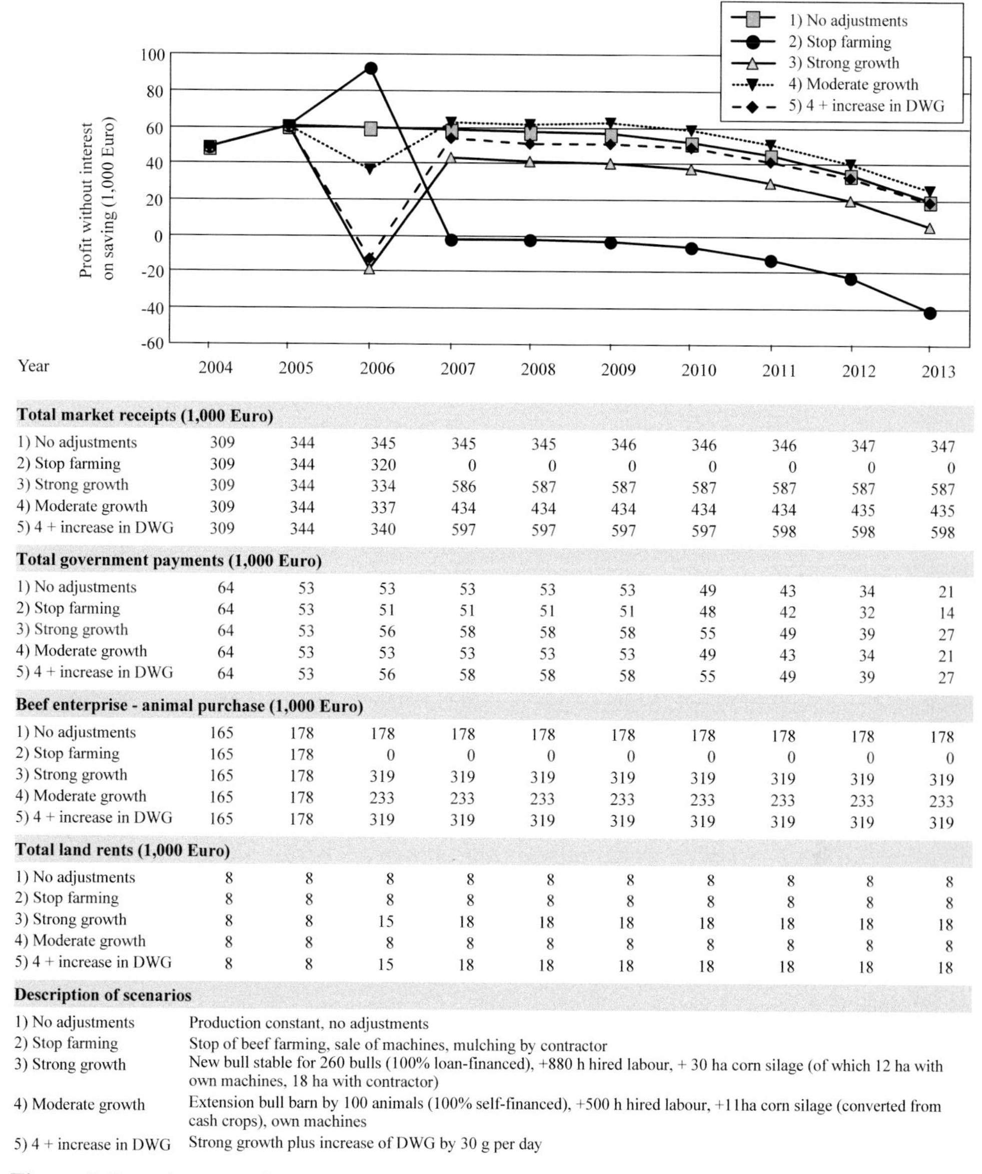

Year	2004	2005	2006	2007	2008	2009	2010	2011	2012	2013
Total market receipts (1,000 Euro)										
1) No adjustments	309	344	345	345	345	346	346	346	347	347
2) Stop farming	309	344	320	0	0	0	0	0	0	0
3) Strong growth	309	344	334	586	587	587	587	587	587	587
4) Moderate growth	309	344	337	434	434	434	434	434	435	435
5) 4 + increase in DWG	309	344	340	597	597	597	597	598	598	598
Total government payments (1,000 Euro)										
1) No adjustments	64	53	53	53	53	53	49	43	34	21
2) Stop farming	64	53	51	51	51	51	48	42	32	14
3) Strong growth	64	53	56	58	58	58	55	49	39	27
4) Moderate growth	64	53	53	53	53	53	49	43	34	21
5) 4 + increase in DWG	64	53	56	58	58	58	55	49	39	27
Beef enterprise - animal purchase (1,000 Euro)										
1) No adjustments	165	178	178	178	178	178	178	178	178	178
2) Stop farming	165	178	0	0	0	0	0	0	0	0
3) Strong growth	165	178	319	319	319	319	319	319	319	319
4) Moderate growth	165	178	233	233	233	233	233	233	233	233
5) 4 + increase in DWG	165	178	319	319	319	319	319	319	319	319
Total land rents (1,000 Euro)										
1) No adjustments	8	8	8	8	8	8	8	8	8	8
2) Stop farming	8	8	8	8	8	8	8	8	8	8
3) Strong growth	8	8	15	18	18	18	18	18	18	18
4) Moderate growth	8	8	8	8	8	8	8	8	8	8
5) 4 + increase in DWG	8	8	15	18	18	18	18	18	18	18

Description of scenarios	
1) No adjustments	Production constant, no adjustments
2) Stop farming	Stop of beef farming, sale of machines, mulching by contractor
3) Strong growth	New bull stable for 260 bulls (100% loan-financed), +880 h hired labour, + 30 ha corn silage (of which 12 ha with own machines, 18 ha with contractor)
4) Moderate growth	Extension bull barn by 100 animals (100% self-financed), +500 h hired labour, +11ha corn silage (converted from cash crops), own machines
5) 4 + increase in DWG	Strong growth plus increase of DWG by 30 g per day

Figure 5. Farm income of a 260 bull finishing farm under different farm strategies.

certain levels of farm incomes. In particular, the analysis focused on the impact that variations in key prices and productivity indicators have on the economic result of the farms when applying the five adjustment strategies. For this analysis, the SIMETAR software developed by Richardson, Texas A&M University, was used.

Definition of key input and output variables (KIV and KOV)

A set of prices and productivity indicators was selected, all of which have a high impact on the profitability of beef finishing. The variables were:
- Beef price;
- Calf price;
- Animal losses (mortality);
- Daily weight gains (converted into final weights);
- Concentrate/grain price.

Further, key output variables (KOV) were defined for the final analysis of the results. The main KOV shown here is the farm income.

Time series analysis

For the variables selected above, ten years of historic time series data were collected. These data were then analysed in a multiple step process with the objective to obtain distribution functions for each variable. The steps undertaken were (see also Richardson, Schumann, 2005):
- Plotting of the time series data to identify specific events and discontinuities; if necessary, manual correction or introduction of a dummy variable.
- Simple or multiple regression to obtain a trend for each KIV (key input variable).
- Performing T- or F-Test for significance.
- Calculation of a correlation-matrix and standard-normal distribution matrix for the KIV.
- Creation of a matrix with correlated standard normal distributions.
- Based on this matrix, Excel calculates correlated standard normal distributions for each variable.
- Finally, the 'Emprical' Function of SIMETAR is used to generate empirical distribution functions of the KIV reflecting the historical variation of the past (Richardson, Schumann, Feldmann, no year).
- Albeit technically possible, no assumptions and adjustments were made about future changes in the variations compared with the historical variations. It can, however, be assumed that price volatility will increase with further liberalisation of markets. This would mean that the risk involved in beef production is going to increase.
- The distribution functions for the KIV were then applied to the deterministic strategy analysis by running the model TIPI-CAL plus the SIMETAR add-in for 100 iterations with randomly and simultaneously selected values for each KIV.

The result is 100 single values for each KOV selected which again is used to calculate confidence intervals and cumulated distribution functions for each strategy analysed.

Results of the risk analysis

Table 2 shows the results of the strategy analysis reflecting risk for the DE-260 farm. Additionally, the results for the Baseline were included to widen the picture. Again, farm income expressed in EURO 1,000 was chosen as the KOV and the last year of simulation (2013) is displayed in the table. The table displays the result of the deterministic model-run already shown in Figure 5 as well as the confidence intervals of the 5 and 95 percentiles as well as the 25 and 75 percentiles. The former indicates the range in which 90 percent of the simulation results would fall in, the latter indicates the range where 50 percent of the simulation values can be found.

Table 2. Deterministic values and confidence intervals for the farm income of the baseline and five adjustment strategies analysed for a 260 bull finishing farm for the final year of simulation (2013) (Keller, 2006).

Strategy	Deterministic 1000 EURO	90% of the values in the range from EUR 1000 … to EUR 1000	50% of the values in the range from EUR 1000 … to EUR 1000
Baseline (Agenda 2000)	42	16 to 63	31 to 53
1) No adjustments	21	-6 to 43	10 to 32
2) Stop farming	-41	n.r.	n.r.
3) Moderate growth	26	-8 to 55	11 to 41
4) Strong growth	4	-44 to 46	-15 to 24
5) = 4 + increase DWG	18	-32 to 59	-2 to 38

n.r. = not relevant

The following conclusions can be drawn:

- The *stop farming* strategy only shows a deterministic value because after termination of production none of the KIVs are relevant anymore in the overall result.
- The *moderate growth* strategy shows the highest mean value (deterministic) and appears to be the best adjustment strategy. The two *strong growth* strategies fall behind, even when comparing them with the no adjustment strategy.
- The higher the cattle number, the higher the variation of the profit and the risk of running into a loss is. This holds true for all growth strategies.

Figure 6 supports the impression obtained from the analysis presented in Figure 5 and Table 2: the moderate growth strategy appears as being the most promising for this farm and under the assumptions made. The figure shows the cumulative density functions for the farm income of all strategies analysed.

- The *stop farming* strategy shows a loss of approximately EURO 41,000 for the whole distribution.
- The *no adjustment* strategy has a probability of generating a loss of less than ten percent but also just a probability of less than five percent of making a profit of more than EURO 40,000.
- The *strong growth* strategy can not generate much higher profit than the no adjustment strategy but it reveals a probability of 45 percent of generating a (huge) loss. The strategy *strong growth plus increase in daily weight gains* shows the highest potential profit but also a probability of 30 percent of creating a loss.

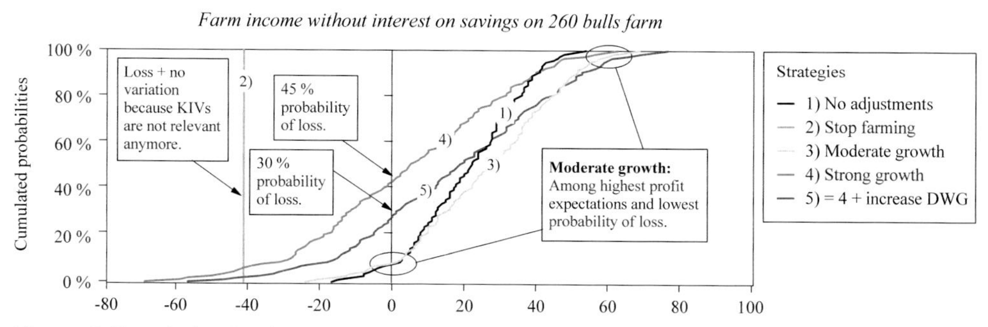

Figure 6. Cumulative density functions for the farm income of the five adjustment strategies analysed for a 260 bull finishing farm for the final year of simulation (2013) (Keller, 2006).

- Finally, the *moderate growth* strategy provides a probability of 50 percent of making a profit beyond EURO 30,000 and at the same time a probability of less than ten percent of making a loss.
- These findings support the view that can be taken from the deterministic analysis described above. They were further supported by the advisors involved in the definition of the strategies.

With one exception, the moderate growth strategy is also among the most promising strategies for the other farms analysed. The exception is the DE-160F which had a sow enterprise that was replaced by the growing beef enterprise in the simulations. For this farm, continuing beef production without changes is the best alternative.

The results presented here should hold true for most rather specialised and reasonably large sized beef finishing farms with comparable production systems and livestock densities in Germany. For significantly smaller farms as well as for mixed farms and less intensive farms separate analysis should be carried out. It can, however, be assumed that in the absence of economies of scale and with less degree of specialisation the economic situation of the beef enterprise will turn out to be even worse than on the farms analysed here.

Conclusions about the risk analysis

If the conclusions from the deterministic and the stochastic analysis are similar, one might indeed ask about the benefits of the risk analysis, in particular if the collection of data for the risk analysis creates efforts and costs. The following conclusions can be drawn:

- If animal numbers between two strategies are similar and the farm income shows significant differences, a risk analysis might not be necessary. In this case, the strategy with the higher farm income is most likely the favourable one.
- The same conclusion applies if the farm incomes are similar with significant differences in animal numbers. In this case, the strategy with the lower number of animals usually is less risky and should be favourable.
- If none of the two cases above is true, a risk analysis should be performed to avoid wrong conclusions.
- In any case, risk analysis quantifies the risk involved and the probability of certain events. It is therefore useful additional information for the decision maker.
- Further, risk analysis provides the option to vary the variability of KIV and their implication for the result. This becomes particularly relevant when increases of future price volatility can be expected due to reductions of tariff rates and fluctuations of beef supply due to climate changes.

References

Agra Europe, 2004. Issue 35/04, EN 1-3.

Agra-Europe, 2004. AgraFacts 30.01.04.

Agra-Europe, 2004. Issue 11/04, EN 12-14.

BMVEL, 2004. Meilensteine der Agrarpolitik – Umsetzung der europäischen Agrarreform in Deutschland. Bundesministerium fuer Verbraucherschutz, Ernährung und Landwirtschaft BMVEL, Berlin.

Brömmer, J,. 2005. Produktionssysteme, räumliche Verteilung und Struktur der Rindermast in Deutschland – eine expertengestützte Analyse. Diploma thesis, University of Applied Sciences Osnabrück.

EU-COM, 2005. Overview of the implementation CAP reform (first and second wave of the reform, reform of the sugar sector) http://europa.eu.int/comm/agriculture/markets/sfp/ms_en.pdf.

Deblitz, C., 2006. Agri benchmark beef report 2006: Benchmarking farming systems worldwide. Braunschweig : FAL, 63 pages.

Deblitz, C., Y. Zimmer, 2005. A standard operating procedure to define typical farms. Unpublished manuscript http://www.agribenchmark.org/methods_typical_farms.html.

DBV (German Farmers Association). Die Reform der Gemeinsamen Agrarpolitik. Bonn.

Keller, M., 2006. Betriebliche Entwicklungsstrategien für ausgewählte Rindermastbetriebe unter Berücksichtigung von Risiko. Master thesis, Georg-August University Göttingen.

Richardson, J. and K. Schumann, 2005. Modelling Correlation of Non-Normal Distributed Random Variables in Stochastic Models. Department of Agricultural Economics, Texas A&M University.

Richardson, J, K. Schumann and P. Feldman. Simetar© - Simulation & Econometrics to Analyse Risk. SIMETAR Handbook

ZMP, 2007. Marktbericht Vieh und Fleisch, 49. Jahrgang, No. 3, from 19.01.07.

ZMP, 2006. Marktbilanz Vieh und Fleisch 2006, Bonn.

Adaptation of Irish beef farming systems to the Luxembourg agreement reform of the Common Agricultural Policy (CAP)

M.G. Keane

Teagasc, Grange Beef Research Centre, Dunsany, Co. Meath, Ireland

Abstract

Total cattle and cow numbers are 6.89 and 2.35m respectively. Beef cows account for 51% of total cows. Considering the country as three regions, namely West, South and East, 40% of the beef cows and 11% of the dairy cows are in the West, whereas 31% of the beef cows and 67% of the dairy cows are in the South. Total slaughterings are about 1.7m head with 0.25m head exported live. Total beef output is about 0.57mt of which 85% is exported. Cattle production systems are diverse with males raised predominantly as steers yielding a mean carcass weight of 355kg. Animals are generally cross-bred with 75% of beef herd calves sired by Charolais and Limousin. About 50% of dairy cows are crossed with beef bulls of both early and late maturing breed types. The reform of the CAP resulted in full decoupling of all premia and substitution of a single farm payment which has strict regulations and restrictions attached. To date, there have been few effects of decoupling on the beef industry and earlier predictions of a large decrease in beef cow numbers and beef output have not come about. It is too early to determine the effects on farm incomes as there was a carry-over of premia into the decoupled era.

Keywords: beef production, CAP reform, decoupling

Irish beef production

Cattle numbers

From 1996 to 2005, dairy cow numbers decreased from 1.266m to 1.117m (12%), beef cows increased from 1.113m to 1.231m (11%), total cows decreased from 2.379m to 2.348m (1%) and total cattle decreased from 7.313m to 6.889m (6%). Most of these changes occurred before 2000 and changes since then have been negligible. Beef cows now account for 51% of total cows.
A total of 2,150,065 calf births were recorded in 2005 (CMMS Report, 2005). The geographical distribution of calves born to beef and dairy cows (representing the geographical distribution of beef and dairy cows) is shown in Figure 1. If the country is considered as three regions, namely West (Donegal to Clare), South (Kerry to Wexford) and East (remainder), beef cows are reasonably evenly distributed across the regions (West 40%, South 31%, East 29%), whereas dairy cows are predominantly in the South (67%) with only 11% in the West. Over 75% of calves were sired by beef bulls. About one third of the beef sired calves came from dairy cows. In the first, second, third and fourth quarters of the year, 41, 42, 10 and 7%, respectively of beef sired calves were born. The corresponding figures for dairy sired calves were 67, 19, 6 and 8%. Overall, almost half of all calves were born in the first quarter, with a further one third born in the second quarter, leaving only one sixth born in the second half of the year.
From the national herd of 2.35m cows, about 1.9m cattle are marketed annually. Of these, 1.5 to 1.8m are processed through Irish slaughter plants and the remainder are exported live. Live exports vary widely. Over the last 10 years the lowest number was 57,000 in 1997 and the highest was 416,000 in 1999. As live exports usually occur in the year of birth whereas slaughter does not occur until later

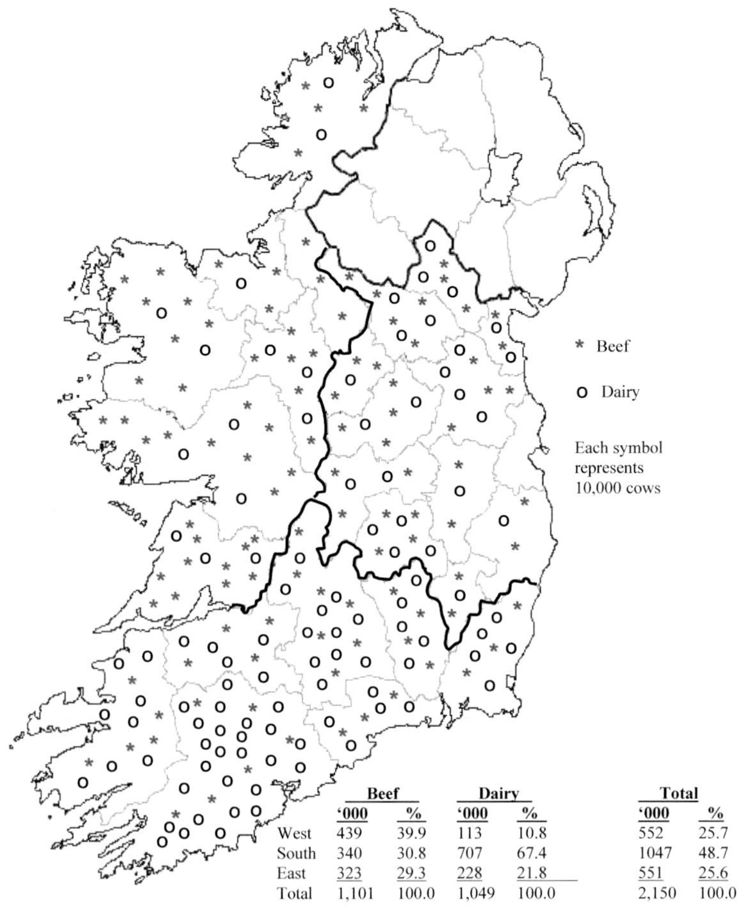

	Beef		Dairy		Total	
	'000	%	'000	%	'000	%
West	439	39.9	113	10.8	552	25.7
South	340	30.8	707	67.4	1047	48.7
East	323	29.3	228	21.8	551	25.6
Total	1,101	100.0	1,049	100.0	2,150	100.0

Figure 1. Distribution of beef and dairy cows in Ireland.

years, adding the live exports to the slaughtered animals for the same year gives a misleading figure for cattle output. For example, as a consequence of the low cattle exports in 1997, there were high slaughterings in 1999 (1.997 m) but there were also high live exports in that year (416,000) resulting in cattle output for 1999 reaching 2.413m. In contrast, cattle output in 1997 was only 1.682m, as a consequence of low slaughtering in that year due to high live exports previously. Excluding cows, steers, heifers and young bulls comprise 63%, 29% and 8% of cattle output, respectively.

Slaughterings and beef output in 2005

Over 1.6m animals were slaughtered in 2005. These comprised 42.2% males from beef sires, 30.3% females from beef sires, 13.9% males from dairy sires and 13.7% females from dairy sires. Few animals were slaughtered before 18 months of age (Table 1). About 30% of beef and dairy males, 40% of beef females and 90% of dairy females were over 30 months at slaughter. The latter were predominantly cull cows.

Table 1. Distribution (%) of slaughterings by age, gender and breed type, 2005.

Age (mts)	<-18	18-24	24-30	30+	Total
Beef sire progeny					
Males	6.6	16.8	43.2	33.4	677,117
Females	7.0	26.7	24.8	41.5	486,284
Dairy sire progeny					
Males	4.9	17.5	47.6	29.9	223,145
Females	0.8	2.0	5.2	91.9	219,448
Total slaughtering					1,605,994

Beef output in 2005 was 574,000 t. Of this, 94% was domestically produced and 6% was imported including imports from Northern Ireland. From this total supply, 15% was used domestically and 85% was exported. The distribution of exports between the UK, Continental EU and international markets was 53%, 39% and 8%, respectively. Within continental EU, similar amounts (8-9%) went to France, Italy, Netherlands and Scandinavia. The main international market was Russia.

Live exports

Live exports in 2005 totalled 186,062. Of these, 66% were the progeny of beef sires and 34% were the progeny of dairy sires. The beef sire progeny were 46% male and 54% female, whereas the dairy sire progeny were 97% male. Over 75% of the dairy sire progeny were exported before 6 weeks of age in the first half of the year and a further 10% were exported from 6 weeks to 6 months of age. These were calves destined for veal production. Between 40% and 50% of the beef sire progeny were exported at 6-12 months of age in the final 4 months of the year with a further 20-25% exported in the period 12-18 months of age. These were weanlings destined for feed lots. Most dairy females were over 24 months of age suggesting they were in-calf, or freshly calved, heifers.
Live exports went predominantly to EU countries with Spain, Italy, Netherlands and Northern Ireland accounting for 27%, 26%, 24% and 15%, respectively. The only non EU market of consequence was Lebanon. Late maturing breed types (Charolais, Limousin and Belgian Blue) accounted for 69%, 95% and 81% of the animals exported to Spain, Italy and Northern Ireland, respectively. The animals exported to the Netherlands were 87% Friesians while those exported outside the EU were 74% Friesians. Only 37% and 47% of the animals exported to Spain and Italy, respectively, were males, whereas 97% and 67% of those exported to Netherlands and Northern Ireland, respectively, were males.

Irish beef farming systems

National patterns

There were a total of 118,291 cattle herds in 2005. Of these, 42, 22, 12, 8 and 16% had <25, 25-49, 50-74, 75-99 and >100 animals, respectively. For the purposes of the National Farm Survey (National Farm Survey - NFS, 2005), farming systems are classified as Dairy, Dairy and Other, Cattle Rearing, Cattle Other, Sheep and Tillage. The percentages of total farms in these categories in 2005 were 16.3, 9.4, 24.6, 26.8, 16.3 and 6.6, respectively. Corresponding average farm sizes (utilisable agricultural area, UAA) were 44, 52, 27, 30 and 39 and 60 ha. With the exception of Dairy and Tillage these systems are not very specialised. All systems have some beef cows and cattle. The two cattle systems (Cattle Rearing and Cattle Other), which comprise 50% of total farms, have 70% of

beef cows, 46% of beef cattle and 42% of total cattle. The two dairy systems (Dairy and Dairy and Other), comprise 25.7% of farms, and have 99% of dairy cows. They also have 11% of beef cows, 42% of beef cattle and 46% of total cattle. Sheep and Tillage farms have 19% of beef cows, 13% of beef cattle and 12% of total cattle.
Many cattle farms, particularly in the Cattle Rearing system, are too small to provide an adequate family farm income. Using 0.75 labour unit requirement as the definition of a full time farm, only 38% of all farms were full time in 2005. This varied with system. For the 6 systems as listed above, 93, 67, 11, 20, 32 and 53%, respectively, were full time. Thus, about 85% of all dairy farms were full time as compared with only 15% for beef farms. Full time Cattle Rearing and Cattle Other farms had 54 and 65 ha UAA and 101 and 136 cattle, respectively. On full time cattle farms, 24% of farmers and 34% of spouses had off farm jobs and 10% had other off farm income. The corresponding figures for part time cattle farms were 54%, 30% and 34%. Full time cattle farms had 33% of the UAA and 39% of the total cattle in the Cattle Rearing and Cattle Other systems. Family farm income (€'000) on full time farms for the 6 systems as listed was 41, 49, 29, 42, 29 and 45, respectively. Corresponding percentages of those incomes contributed by direct payments were 48, 73, 122, 117, 121 and 97.

Cattle production systems

There is a wide diversity of cattle production systems and cattle are widely traded at intermediate stages of production. Farmers do not always adhere rigidly to specific systems but may switch between systems in anticipation of improved returns. As the vast majority of calves are spring-born, production systems are mainly based on spring-born calves. There is essentially no veal production and all dairy calves not exported live are raised for beef. Bulls comprise a small proportion of total cattle and are often exported live before finishing. About 50% of dairy cows are crossed with beef bulls so the dairy calves are approximately 50:50 pure dairy and beef x dairy. Females of the latter are often retained as suckler cows. Taking account of these industry characteristics, 4 beef cow based, and 10 non beef cow based systems are recognised in the Management Data for Farm Planning Manual (2005). These systems result from some producers rearing animals from birth to intermediate stages of production and then selling to others who rear them to a later stage of production or to slaughter. Most transfers of animals between farms occur in spring and autumn coinciding with turn-out to pasture and housing. Thus, systems tend to be about 6 months or multiples of 6 months in duration.
The on-farm performance monitoring programme of the Teagasc advisory service (Teagasc e-Profit Monitor Analysis, Drystock Farms 2005), categorises cattle production into 3 systems, namely (i) calf to beef: beef cow farms rearing the progeny to an advanced stage or to slaughter, (ii) calf to weanling/store: beef cow farms selling progeny at 9 to 18 months, (iii) non breeding: beef farms without beef cows. The monitored farms are in the top 25% of cattle farms when benchmarked against the NFS sample. In 2005, 82, 73 and 73 farms were monitored in the categories as listed above. Average farm size (UAA) was 60, 40 and 48 ha, respectively. Stocking rate was 1.84, 1.71 and 1.70 livestock units (LU) per ha and live weight produced per ha was 654, 518 and 690 kg, respectively. Average gross margin was €408/ha, but with fixed costs of €416/ha, profit in the absence of direct payments was negative. There was little difference between the systems in financial outcome which was best for calf-to-beef and poorest for calf to weanling/store.
The farms were divided into the top, middle and bottom one third categories based on gross margin per ha. Compared with the bottom category, the top category had a greater farm size, a 25% higher stocking rate and 53% higher live weight output per ha. The value of output was 2.5 times higher for the top category whereas variable and fixed costs were only 37% and 57%, respectively higher. The low unit price of the output for the bottom category suggested they were making a loss trading in cattle. Overall, the top category had a net profit of about €200/ha and retained 130% of their direct payments whereas the bottom category had a net loss of about €200/ha and only retained 70% of their

direct payments. Total direct payments differed little between the top and bottom categories (€753 and €692) indicating that the higher profitability of the top category resulted from better technical performance and more efficient use of inputs (more output value per unit input) rather than from higher direct payments.

Evolution of beef farming

Herd size and specialisation

Across all farms, average Irish farm size is about 35 ha which (Census of Agriculture, 2000) compares with an EU-15 average of about 20 ha and an EU-25 average of about 15 ha. Average farm size is increasing very slowly by about 0.6 ha per annum. This is because land mobility is low and land price is high. Annual land sales have fallen from about 15,000 ha 10 years ago to about 5,000 ha at present. Over the same period land price has more than trebled. Between 1991 and 2000 (the last year for which there is a complete agricultural census), dairy farm numbers declined by 4% per annum whereas the number of beef farms remained static. There is no evidence that this pattern has changed.

Projecting farm numbers

Irish farms can be classified on the basis of viability into two broad groups, economically viable and non viable (Table 2). An economically viable farm is defined as having (a) the capacity to remunerate family labour at the average agricultural wage, and (b) the capacity to provide an additional 5% return on non-land assets. Non viable farms are not economically viable as a business, but where the farmer or spouse have an off farm job, the household may be sustainable (Agri Vision 2015, 2004).
In 2002, 28% of farms were viable, 27% were part time and 45% were transitional. The average annual decrease in farm numbers is 2%, but this varies greatly between categories. Considerably more farmers are likely to exit dairying as a consequence of the Luxembourg Agreement than would have exited under a continuation of past trends. However, these will not necessarily exit farming but will become part time farmers that are economically sustainable. Thus, the total number and proportion of economically sustainable farms will increase.
It is anticipated that over 40% of existing farmers (in 2002) will retire by 2015. In 2002, 20% of farmers' heirs indicated they would become full time farmers, 73% indicated they would be part time farmers and 7% planned to exit farming.

Table 2. Farm categories.

Viable
Labour supplied is >0.75 labour units and family farm income is sufficient to cover family labour and give a return on assets.

Non-viable
Labour supplied is <0.75 labour units and family farm income is insufficient to cover family labour and return on assets.

Part time:	Economically non-viable but the farmer and/or spouse has an off-farm job leaving the household economically sustainable.
Transitional:	Economically non viable and there is no off-farm employment: not sustainable in the longer term.

Agri Vision 2015

By 2015, the 105,000 farms then remaining will comprise of 40,000 viable farms on three-quarters of which either the farmer or spouse will have an off-farm job. There will be 45,000 part time economically sustainable farms and 20,000 transitional farms (Table 3). Assuming that 75% of non dairy farms are cattle farms, the breakdown of the 40,000 viable farms in 2015 will be 12,500 dairy, 21,000 cattle and 6,500 other. In addition, there will be 32,500 part time sustainable cattle farms and 14,000 transitional cattle farms. Thus, by 2015 a considerable proportion of the land and cattle resources devoted to cattle production will be in part time and transitional farms.
In summary (Table 4), by 2015 about 15% of total farms will be dairy farms, and two thirds will be beef farms. Of the latter, about 30% will be viable, 50% will be part time sustainable and 20% will be transitional.

Implementation of CAP reform in Ireland

Single payment scheme

Following publication of the EU Council Regulation 1782/2003 in September 2003, livestock premia were decoupled from production with effect from 1st January, 2005 and were replaced by a new single payment scheme (SPS) (The Single Payment Scheme, 2004). The payment under the scheme includes all the decoupled payments to which the farm is entitled and is known as the Single Farm Payment (SFP). It commenced on 1st December, 2005 and 2006 was the first full year of its operation.
The cattle premium schemes decoupled from production were:

- special beef premium - young bulls, 1st and 2nd age steers;
- suckler cow premium;

Table 3. Farm population: 2002 v. 2015.

Farm Group	2002		2015[1]	
	No.	%	No.	%
Viable (of which are dairy)[2]	39,000 (-)	29	40,000 (12,500)	38
Part-time sustainable (of which are dairy)	37,000 (-)	27	45,000 (1,500)	43
Transitional (of which are dairy)[2]	60,000 (-)	44	20,000 (1,500)	19
All farms (of which are dairy)[2]	136,000 (-)		105,000 (15,500)	

[1]Past trends adjusted for effects of Luxembourg Agreement.
[2]Not estimated for 2002.
Agri Vision 2015

Table 4. Distribution of Irish farms by system in 2015[1].

	Viable		Part time[2]		Transitional		Total	
	No.	%	No.	%	No.	%	No.	%
Dairy	12,500	80.6	1,500	9.7	1,500	9.7	15,500	14.7
Beef	21,000	31.1	32,500	48.2	14,000	20.7	67,500	64.3
Other	6,500	29.5	11,000	50.0	4,500	20.5	22,000	21.0
Total	40,000	38.1	45,000	42.9	20,000	19.1	105,000	100.0

[1]Past trends adjusted for effects of Luxembourg Agreement.
[2]Sustainable.
Agri Vision 2015

- slaughter premium;
- extensification premium;
- national envelope top-ups for dry heifers, calved heifers and slaughtered heifers.

Generally, the SPS is applicable to farmers who actively farmed during the reference years of 2000, 2001 and 2002 and were paid livestock or arable aid premia in one or more of those years, and who continued to farm in 2005. Entitlements are not attached to land, they are attached to the farmer who established them. The entitlements are calculated based on the average number of animals on which payment was made under each scheme in the reference period multiplied by the 2002 rate of payment for each scheme. The sum of these amounts is the gross single annual payment. This value is then divided by the average number of ha farmed in the reference period to give the number of entitlements and the monetary value of each entitlement.

National reserve

Each Member State was obliged to create a National Reserve, using initially up to 3% of the value of entitlements established by farmers who were farming during the reference period. Certain categories of farmers (including those who commenced farming after the reference period) can be allocated these entitlements.

The National Reserve is allocated amongst the following categories:

- Farmers who inherited a farm from another farmer who died during the reference period and had the land leased at that time.
- Farmers who bought a farm during the reference period which had been leased at that time.
- Farmers who entered a long term lease of a farm during the reference period where the conditions of the lease cannot be altered.
- Farmers who made investments or purchased land during the reference period to increase production.
- Farmers who participated in national programmes of reconversion of production during the reference period.
- People who for certain reasons could not activate their full entitlements in 2005 can do so in 2006 and 2007 but if they are not used for a period of 5 years they are transferred to the National Reserve.

The National Reserve is replenished by

- Up to 50% of the value (or number) of entitlements sold without land in the first 3 years and up to 30% afterwards.
- Up to 10% of the value (or number) of entitlements sold with land.
- Up to 5% of the value (or number) of entitlements where the farm is sold.
- Unused entitlements.

Modulation

Modulation is the process whereby there is a mandatory reduction of each farmer's SFP by a predetermined percentage (3% in 2005, 4% in 2006, 5% in 2007). National Governments can opt for a greater reduction. The first €5000 of entitlements is exempt from modulation. Up to 84% of the funds generated through modulation (ca. €34 m in 2007) can be retained in Ireland for spending on rural development measures. Currently, these measures are the Disadvantaged Areas Compensatory Allowance Scheme, the Early Retirement (from farming) Scheme, the Rural Environment Protection Scheme (REPS), the Forestry Premium Scheme and the proposed Beef Herd Welfare and Quality Improvement Scheme.

Cross compliance

Cross compliance has two key elements:
i. Farmers must comply with a number of statutory management requirements set down in EU legislation on the environment, food safety, animal health and welfare, and plant health.
ii. The farm must be maintained in good agricultural and environmental conditions.
There is also an obligation to ensure that there is no significant reduction in the area of land in permanent pasture compared with 2003.

Force majeure

Force majeure is defined as circumstances outside the farmer's control which could not have been foreseen, and which he, as a prudent farmer, took reasonable precautions to avoid. They include:
- Death of the farmer.
- Long term professional incapacity of the farmer.
- A natural disaster gravely affecting the land.
- Accidental destruction of livestock buildings.
- An epizootic disease affecting the livestock.

In these circumstances, entitlements are based on years (1997-1999) other than the reference years (2000-2002).

Consolidation of entitlements (stacking)

Individual farmers may have difficulty in always having the same land area as they had during the reference period. Accordingly, there is provision (Article 42.5 of Council Regulation No. 1782/2003 and Article 7 of Commission Regulation No. 795/2004) to use the National Reserve to consolidate payment entitlements on the land area currently farmed. This entails surrendering the original entitlements to the National Reserve in exchange for a lower number of entitlements with a higher unit value. In this way the total single payment remains the same.
To consolidate, the farmer must declare the total area currently available for farming and this must be equal to at least 50% of the average area declared during the reference period. Consolidation is not available to farmers who declare fewer ha than entitlements due to the disposal of land by sale or lease other than to a public authority for non agricultural use. The consolidation provisions apply to the following categories of farmers:
- Those who have afforested some of their land since the beginning of the reference period.
- Those who have disposed of land to a public authority for non agricultural use.
- Those who had land leased/rented-in during the reference period but the lease/rental agreement has now expired.
- Those who declared lands in Northern Ireland in the reference period.

Farmers' intentions

In Autumn of 2003 and 2004, surveys were conducted on random samples of farmers in association with the NFS. The objectives were (i) to determine their awareness of CAP reform and decoupling, (ii) to determine their response to decoupling in respect of their farm enterprises, and (iii) to ascertain their views on the impact on farm incomes and farm inputs.
A high proportion (85%) of farmers were in favour of full decoupling with 11% preferring partial decoupling. This was consistent across farming systems. There was a high awareness (81%) of the forthcoming SPS with again little difference between farming systems. The average estimated SFP

ranged from €6,100 for specialist dairy farms to €19,000 for tillage farms. It averaged €6,900 and €11,000 for the Cattle Rearing and Cattle Other systems, respectively. Correspondingly, the per ha payment ranged from €146 for Dairy to €363 for Tillage. It averaged €254 and €356 for the Cattle Rearing and Cattle Other systems, respectively. In terms of enterprise size, 88% did not plan to change the area farmed, 8% expressed the intention to decrease and only 3% expected to increase. The latter tended to be Dairy and Tillage specialists and almost all farmers (96%) intended to use their entitlements. Few intended to sell or buy entitlements.
The expected effect on farm income by 2010 is shown in Figure 2. On average, 41% of cattle farmers expected no change, 27% didn't know what the effect might be, 21% expected a decrease and only 11% expected an increase. In terms of purchased farm inputs, 59% of cattle farmers expected no change, 29% expected a decrease, 8% didn't know and only 3% expected an increase. A total of 8% of dairy farmers intended to exit dairying. These were mainly herd sizes of < 20 cows.

Cattle enterprise post decoupling

Of farmers with cattle, only 4% intended to exit the cattle enterprise, mainly those engaged in both dairying and beef production, but 21%, again mainly those also involved in dairying, expressed an intention to change their cattle system. Over two thirds of cattle farmers did not intend to change the quality of their stock but the other one third intended to improve quality. In terms of cattle numbers, nearly half of all cattle farmers did not intend to change their number of LU while over one third intended to reduce LU. When the intended increases and reductions were aggregated, the intended decrease in LU exceeded the increase by a ratio of 5:1.
Farmers were also asked about their plans to join the REPS. In 2004, 45,000 farms (40%) were already in REPS. While some existing farmers in REPS did not plan to continue, a further 20% aspired to join for the first time.
These findings of farmers' intentions in 2003-2004 should be interpreted as indicative rather than absolute because both the policy and the farmers' familiarity with it were evolving at the time.
FAPRI–Ireland (2003) modelled the likely effect of decoupling. In the absence of any reform, dairy farm numbers would decline by about 15% from 2002 to 2012. With decoupling, the estimated decline was 35%. Thus, the 28,000 dairy farms in business in 2002 were expected to fall to 15,000 by 2012 (the Agri Vision figure is 12,500 for 2015). A corresponding increase in milk quota per farm was predicted.
In 2002, about 47% of beef farmers had an off-farm job. This would have increased to 50-55% by 2012 in the absence of decoupling but with decoupling was projected to increase to 60-65%. However,

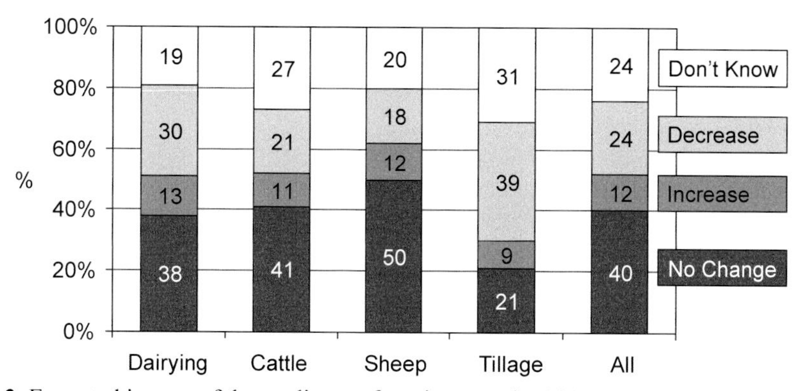

Figure 2. Expected impact of decoupling on farm incomes by 2010.

despite the deterioration in returns from cattle farming relative to off-farm work, a complete shift to part-time cattle farming would not occur because of the farmer age structure. Even though retirement and succession are factored into the projections, the number of farmers aged 60+ will increase from 23% in 2002 to >30% in 2012. Farmers are less likely to work off-farm when they are over 60 but they are also unlikely to retire from farming. Farmers over 60 can work for longer on the farm than off it, and this will curtail the increase in part-time farming and delay farm restructuring.
With the advent of decoupling, various surveys predicted that about 10% of cattle farmers would engage in 'sofa farming' (i.e. they would not farm but would retain their farm to draw the SFP). This percentage was expected to decline as 2012 approached.

Impact of CAP reform on economic results

As 2005 was the first full year of decoupling, there was a large carry-over of direct payments from the previous year as shown by the NFS. This invalidates income comparisons between 2005 and preceding years. Compared with 2004, total output value, including direct payments, increased by 55% in 2005. This was due entirely to an increase of 75% in the direct payments as total livestock output decreased by 12%, comprised of decreases in dairy, sheep and cattle output of 5%, 6%, and 1%, respectively. Both direct and fixed costs increased by 5% each.
In respect of the two main cattle systems, Cattle Rearing and Cattle Other, the value of gross output increased by 33% and 40%, respectively, resulting in corresponding family farm income increases of 75% and 110%. However, much of these increases was due to the carry-over of premia payments from 2004. When incomes are adjusted for these payment overlaps, the increases for the Cattle Rearing and Cattle Other systems were 10% and 25%, respectively. Direct costs decreased by 15% and 11% for the systems, respectively, but fixed costs increased by 11% and 12%. The income increase from 2004 to 2005 could not be attributed to increases in cattle prices, which for steers, heifers and cull cows averaged 4%, 5% and 6%, respectively. Compared to 2005, cattle prices in 2006 were higher by 7% for steers, 6% for heifers and 11% for cows.
Unlike the NFS, the Teagasc e-Profit Monitor Analysis partitions the direct payments to their due year rather than the year in which they are paid. Thus, it is valid to compare a matched sample of 122 farms for 2004 and 2005. Stocking rate and output were unchanged between the years, but the value of output was 17% higher in 2005, which was much greater than the general increase (5%) in cattle price. Perhaps the animals were of better quality or they were sold when the market was relatively more favourable. Variable costs increased by 4% and fixed costs decreased by 8%. These are opposite to the findings of the NFS. Gross margin increased by 37% and the net loss in the enterprise was reduced from €163/ha to €15/ha. Premia were down by about 7% and the proportion of premia retained as profit increased from 82% to 96%.

Consequences for production systems

Societal issues

As affluence increases, society becomes more aware of the external costs involved in food production, and it also can afford to place a higher value on public goods like food safety, animal welfare, and environmental and ethical issues. This intensifies the potential conflict between degradation of public goods and increasing farming efficiency and productivity. The shifts in EU policy over the last decade are indicative of the relationship between commodity output for private gain and supply of public goods. A hypothetical physical relationship between inputs and outputs for both commodity food production and public goods is outlined schematically in Figure 3. The curve for agricultural products (AP) conforms to the standard format with output initially increasing rapidly with each added input,

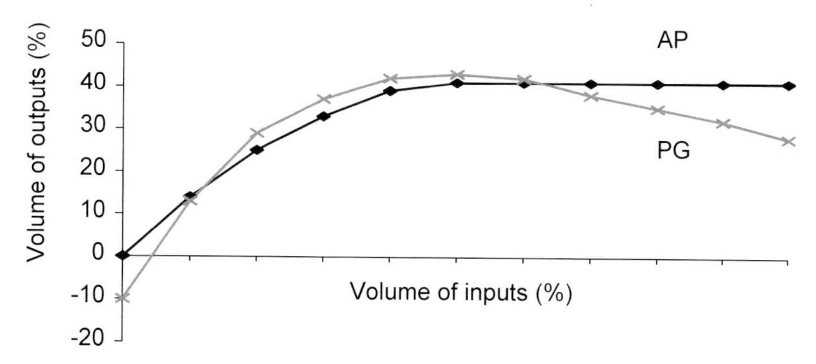

Figure 3. Production function relationship of agricultural products (AP) and public goods (PG) (Dunne, W. 2004).

followed by a slower rate of response as intensification increases, and eventually reaching a plateau, assuming maximum exploitation of existing technology and management skills.
The curve for public goods (PG), in contrast, can be negative at zero or very low levels of inputs. For example, animals may starve due to insufficient winter feed or die due to diseases, public access to certain areas may be restricted by uncontrolled vegetation or there may be abandonment of land. In these situations very small increases in inputs can result in a substantial increase in public goods. For example, the introduction of grazing animals could provide a beneficial level of vegetation control, physical access and a desired type of landscape. Similarly, disease control and/or winter feed could result in significant improvements in animal welfare and even public health benefits. The public goods curve is like an inverted U, initially rising rapidly and then tapering off, followed by a small plateau and then entering a phase of decline. The latter phase represents the high public costs of resource depletion at high levels of farming intensity.
Since the profiles of the two curves differ, as intensification of farming occurs, there are a number of points of intersection. At low levels of farming intensity the two curves are highly complementary and the public gains in resource conservation and development could even exceed the private agricultural gains. After the first point of intersection the curves still remain complementary with largely similar rates of increase, extra inputs continue to add gains for the farmer but also some public goods. As intensification increases, the gain in public goods stalls before entering a relatively rapid decline due to increasing public costs (e.g. serious pollution problems). When this level of intensification is reached, there is antagonism between the private gains of the farmer and the loss of public goods to society.

Production issues

The agricultural products and public goods relationship shown in Figure 3, illustrates the differing responses to resource use. These same production functions are reproduced in Figure 4, but also included are the output to input price ratios before (including direct payments) and after (excluding direct payments) decoupling.
When the value of the direct payments is excluded from the price of the agricultural output, the output to input price ratio declines sharply. In response, the optimum economic level of agricultural output also declines. This leads to a change in the farmers' incentive to supply public goods as demonstrated by the corresponding point on the public goods production function. The economic optimum mix of agricultural products and public goods is very different in the two situations (i.e. before and after decoupling).

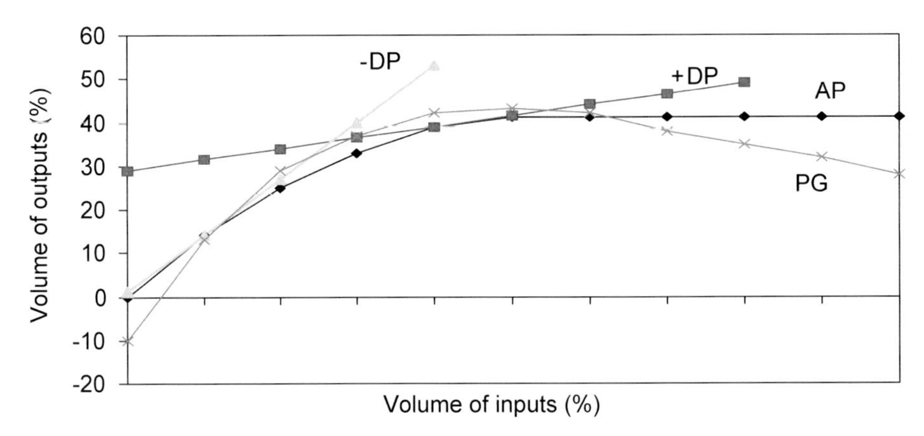

Figure 4. Production function relationships of agricultural products (AP) and public goods (PG) together with the output to input price ratios before (+direct payments (DP)) and after (-DP) decoupling (Dunne, W. 2004).

The magnitude of the changes in the balance of private and public gains and losses depends on the scale of the shifts that occur in the respective production functions. If the economic optimum shifts from very intensive farming to extensive low input farming the public gain may be substantial. Yet, the farmers' return may have altered little when the value of the decoupled payment is taken into account.

Once direct payments are decoupled it is difficult to predict changes in farm product prices, animal numbers and farm inputs, but it is generally accepted that expenditure on concentrate feed, fertilisers and agro-chemicals will decrease in line with the decline in animal numbers and profitability. From a natural resource use and food marketing perspective, this should facilitate the use of more diverse production methods and a move towards more eco-friendly farming and farm products. This in turn should encourage the development of alternative enterprises and products, and facilitate farmer branding by specific locality or region.

Environment and energy issues

Methane accounts for about 16% of total greenhouse gas (GHG) emissions in Ireland, and given that Ireland currently produces substantially more GHG than its Kyoto target, all sectors of the economy, including agriculture, will come under pressure to reduce emissions. As cattle production is the predominant farming system in Irish agriculture, it will be a particular target for emissions reduction. Policy issues such as CAP reform and the Nitrates Directive will result in some reduction in Irish cattle numbers as mentioned elsewhere. The Irish Government climate change strategy document (Ireland's Pathway to Kyoto Compliance) published in July 2006, estimates that the full effects of CAP reform (Luxembourg Agreement) will be a reduction of 2.4 mt CO_2 equivalent per year from 2008-2012. This will be due to reductions in both methane and nitrous oxide and is a very substantial proportion of total agriculture emissions (19.9 mt CO_2 equivalent in 2004).

With the growing emphasis on replacement of fossil fuels by biofuel, there will be increasing competition between animal production and biofuel production for the world supply of grain. Use of grain in Irish beef production has increased in recent years but this is unlikely to continue as grain prices increase. Margins in beef production are already very low or negative so the industry could not compete for higher priced grain. Technically, the industry could move to production systems that are almost entirely grass based with little grain input but this would have two negative consequences, namely (i) increased methane emissions (due to longer production cycles and more forage in the diet) and (ii) seasonality of production. The former would be unwelcome to the Government and to society generally (even though the methane emissions may be offset to some extent by the sequestration of CO_2 by grassland), while the latter would be unwelcome to the meat processing industry that requires regularity of supply to maintain market share and utilise plant capacity efficiently. To date, neither policy makers nor industry leaders have addressed these potentially critical issues.

Farmer issues

Under the reformed CAP, farmers have much greater farming flexibility which should increase market orientation. For Irish beef farmers, whose production systems were constrained over the years by the CAP regulations, this is a radical departure. Within broad limits, it allows them to change the enterprise mix, vary the methods and intensity of farming, and adjust farming systems to changes in costs and value of output. However, there is little evidence to date of any major changes by beef farmers probably because of the low or negative margins from beef production independent of direct payments. Thus, the putative advantages of CAP reform flexibility have not been realised.
Irrespective of the effects of decoupling, the fortunes of Irish beef farming will continue to be largely determined outside Ireland because of the high proportion of beef exported. At present, it is not possible to discern differential effects of decoupling for different production systems. The special beef premium (SBP) was responsible for limiting, and in some cases reversing, the development of specialised winter fattening (because of the to 180 head per year limit). Now that the SBP has been decoupled, an increase in specialised winter fattening might be again expected. However, the advent of the Nitrates Directive and associated environmental regulations makes the development of large scale winter fattening units more problematic than before. Furthermore, winter fattening is expensive and financially risky as it depends on an unreliable autumn to spring price increase to be profitable. Irish beef plants occasionally subsidise spring beef price from autumn profits in order to hold market share but there is no guarantee they will continue to do this.
As shown earlier, over 50% of Irish beef exports go to the UK. With the return of indigenous beef from over 30 months old animals to the UK market, lower imports from Ireland should be required. If this happens, a market for the displaced Irish beef will have to be found elsewhere. For the present, Irish beef is holding its share of the UK market because of concerns over the traceability of South American beef.
When the impact of the proposed decoupling on Irish beef production was modelled by FAPRI-Ireland, beef cow numbers were predicted to fall by about 20% from 2005 to 2012 as a result of falling beef prices (consequent on some de-stocking following decoupling). This has not happened. There has been no de-stocking and prices have increased rather than decreased. Weanling price is a major factor influencing the profitability of suckler cow systems, and even though the proportion of total weanlings exported is small, the high prices paid by Spanish and Italian importers has greatly supported the weanling market in recent years. If this continues, there is unlikely to be any significant reduction in Irish suckler cow numbers in the immediate future.
By 2012, dairy cows are projected to be about 100,000 below their 2005 number as a result of increasing milk yield. At present, about 50,000 male dairy calves are exported annually to continental EU countries (mainly Netherlands) for veal production. This helps to keep a floor under dairy calf prices. There is a risk that if dairy calf prices were to fall below a certain threshold they would be slaughtered at birth as happens routinely in New Zealand. As about 17% of all calves used for beef production are pure bred dairy males and 33% are pure or cross-bred dairy males, any significant slaughter of dairy calves at birth would have major consequences for calf supply to the beef industry.

Conclusions

As a result of decoupling, the overall rate of decline in farm numbers will not change greatly but the number of dairy farms, which were already declining more rapidly than farms generally, may now decline even more rapidly. As some of those exiting dairying will enter beef production there may be no decrease in the number of beef farms up to 2015. However, only about 30% of those beef farms will be viable, leaving large proportions of the land and animal resources devoted to beef production on part time and transitional farms. This will have consequences for the application of technology, commercial decision making and responsiveness to market signals within the industry. There is little indication that there will be large structural changes or consolidation in the industry or a rapid increase in average farm or herd size.
Dairy cow numbers will continue to decline at the current rate (~1% per annum) but the increase in beef cow numbers in recent years (~1% per annum) is unlikely to be sustained. Instead, beef cow numbers are likely to decrease slightly but the large decrease forecast by FAPRI-Ireland is unlikely to materialise. Thus, assuming that live exports continue at current levels, there will be a small decrease in overall Irish beef output.
The average beef enterprise is unlikely to make a net profit independent of the SFP and other direct payments in the immediate future. At present, direct payments are >120% of family farm income, and only a very small proportion of enterprises have a positive net margin which is generally <€200/ha. While the SFP will decline due to modulation, the total direct payment to beef farmers may not decline because of increasing numbers joining REPS and the proposed new Beef Herd Welfare and Quality Improvement Scheme.
There is no indication that CAP reform favours one type of production system over others, and in any event, the widespread practice of inter farm trading will quickly eliminate any differentials between systems that do arise. Other than between breeding and non-breeding systems, farmers can move readily between systems based on anticipated profitability.
Intuitively, the decoupling model applying in Ireland should favour more extensive systems, but Irish beef systems are already extensive and there are other considerations, particularly for part time farmers, such as the number of generations of animals on the farm, housing facilities and capital tied up in livestock.
More important than CAP reform in favouring some production systems over others will be the extent of the live export trade. Many Irish beef systems are based on the purchase of dairy calves, dairy or suckled weanlings, or stores. High export prices for dairy calves or suckled weanlings or both, coupled with slightly declining dairy and beef cow numbers, should favour breeding systems relative to non breeding systems. However, any large scale movement from non breeding to breeding systems is unlikely in the context of widespread part time farming and existing farmer skills and farm facilities. The effects of the recent Nitrates Directive and its attendant environmental regulations on beef production have yet to elaborated.

References

Central Statistics Office, Census of Agriculture 2000, Ireland. Main results, pp 1-33.

Department of Agriculture and Food, 2004. Agri Vision 2015, Appendix 4.

Department of Agriculture and Food, 2004. The Single Payment Scheme – An Explanatory Guide, p 32.

Department of Agriculture and Food, 2005. CMMS Statistics Report, p 68.

Dunne, W., 2004. Economic consequences of market globalisation on livestock farming systems in Western and Eastern Europe. Proceedings 55th Annual Meeting of European Association for Animal Production, Bled, Paper LI.1.

Teagasc, 2003. The Luxembourg CAP Reform Agreement: analysis of the impact on EU and Irish agriculture, FAPRI–Ireland, p 91.

Teagasc, 2005. Management Data for Farm Planning, pp 11-45.
Teagasc, 2005. National Farm Survey 2005, p 101.
Teagasc, 2005. Teagasc e-profit Monitor Analysis of Dry Stock Farms, p 28.

Impact of the new CAP on Italian beef farming systems

K. de Roest and C. Montanari

Research Center for Animal Production (CRPA), Reggio Emilia, Italy

Abstract

Most of the beef fattening farms in Italy are highly specialised in beef production. The utilised agricultural area of these farms is dedicated by more than 90% to feed production, either for roughage (maize silage) or for concentrates (cereals). This strategy is pursued in order to reach the highest independence possible of the feed market. For this reason, the decoupling of premia does not open a strategy of diversification, as the high rate of specialisation remains the main objective of fattening farms. Feed and weaner prices determine the major part of profitability of the beef farms. The future of beef fattening heavily depends on the development of these prices. Effective application of the Nitrate Directive will take place in 2007 and 2008. The Nitrate Vulnerable Zones (NVZ) have been extended substantially in Veneto, Lombardia and Piedmont. Many fattening farms in Veneto will be located in the NVZs, where a maximum of 170 kg N of manure can be spread. Investment in extra manure storage capacity and transport or treatment of excess manure will increase production costs for fattening farms located in NVZs. The large farms enjoying substantial Single Farm Payments will be able to face the Nitrate Directive problem with relatively less financial efforts than the small farms. However, eventual decisions about maximum ceilings of CAP payments might affect these farms most. It is expected that modulation of payments will generate only limited reductions in subsidies. Of course, a lot depends on the contents of the second reform of the CAP to be carried out in 2008. Beef consumption in Italy will probably decline slightly due to the strong competition of lower priced poultry and pig meat. Meat imports from South America will increase, in particular when a further liberalisation of trade will be decided within the Doha round. Actually this meat is marketed in the restaurant and catering circuits, but in the next years this beef will appear on the supermarket shelves as well. The final result of the abovementioned developments will be that the beef fattening activity in Italy will continue, but at a lower level than the actual one, as the sector has to absorb the extra costs related to the application of the Nitrate Directive and has to face a more fierce competition from abroad. A further intensification strategy is not foreseen as the maize silage system will remain the most convenient way of feeding. The suckler cow system in Italy will continue to feed a niche market of high priced quality beef, based on the local beef breeds of Central Italy. Expansion of this system is not probable, as the natural and climatic conditions pose limits to its growth. For the same reason, the domestic production of weaners of these breeds will not grow significantly, as Italy does not have a comparative cost advantage in this production. Scarce rainfall in summer, low roughage production and a highly fragmented farm structure are not ideal conditions for a profitable production of weaners which can compete with weaners imported from abroad.

Keywords: beef fattening in Italy, production costs, CAP reform, Nitrate Directive, prospects

Beef production in Italy

Italian beef production (1.109 million tons in 2006) consists of approximately 75% red meat from young bulls and heifers slaughtered at an average age of 18-20 months, the remainder being veal calves (13%) and cull cows (12%). In 2006, 2 million 551 thousand animals, young bulls and heifers, were slaughtered at an average live weight of approximately 600 kg. More than 40% of young bulls

and heifers fattened in Italy are derived from the dairy herd which supplies approximately 1 million heads to be finished for meat production. Cattle from the national suckler cow farms represent a minority share and with approximately 480 thousand animals they make up around 15% of all the slaughterings of young bulls and heifers. The remaining amount corresponding to 45% are weaners imported from abroad to be finished in specialised beef finishing farms (Figure 1).
The relatively small dairy herd and the very limited number of suckler cows explain the structural deficiencies of the Italian beef production system, which strongly depends on the foreign supplies of live animals (weaners) and meat. The production of animals born in Italy meets on average only 64% of internal consumption requirements. The shortage of beef cattle increased over the last few years due to the gradual reduction of the domestic dairy and suckler cow herd. In the last ten years, the number of dairy cows decreased by 30%, reaching 1,84 millions of animals following the application of the milk quotas (Figure 2). Italy raises only 472 thousand suckler cows raised in

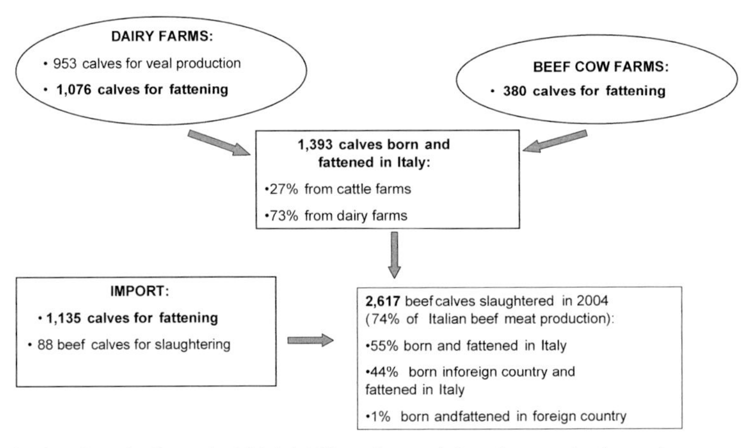

Figure 1. The beef cattle flows in 2004 (.000 males and females aged >1 year).

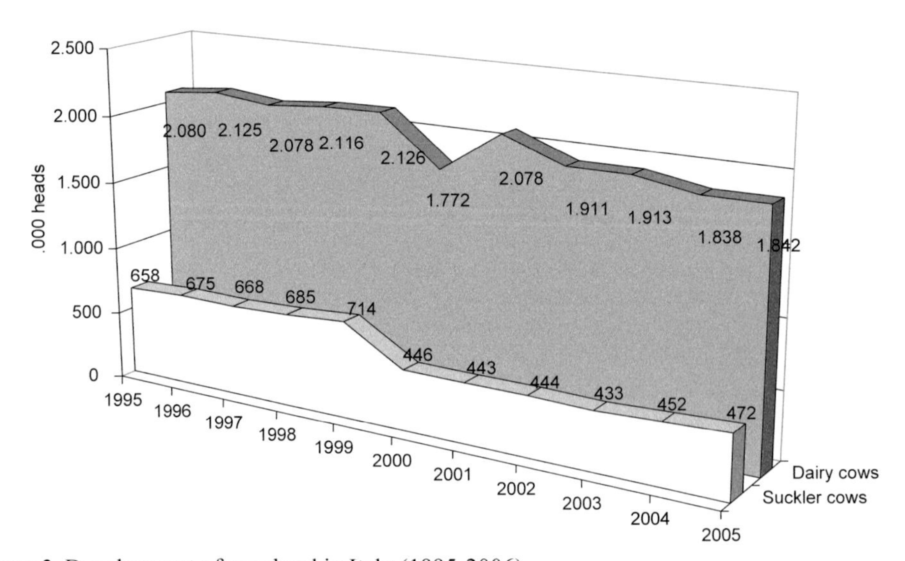

Figure 2. Development of cow herd in Italy (1995-2006).

Piedmont and in the central regions of the country. During the same period, imports of beef and in particular live animals increased steadily at a yearly average rate of 2.5%. Over the last four years, Italy imported on average almost one and a half million live bovine animals per year. One million of these calves are imported within a weight range of 160 to 400 kg destined to be fattened for red meat production, 240 thousand calves are purchased by specialised veal calf producers and 160 thousand heads are imported ready for slaughter.

The Italian beef farming systems

Beef production is characterised by a strong concentration in Northern Italy (Table 1), reflecting the dairy herd concentration within the Po Valley and the excellent suitability for the production of maize silage and other intensive forage crops. The first four regions of Northern Italy – Veneto, Lombardia, Emilia-Romagna and Piemonte – represent two thirds of domestic beef production. Except for Piemonte that boasts a suckler cow herd equal to 23% of the overall suckler cow herd in Italy, the beef farms of the Po Valley are specialised in the fattening of weaners. Against a suckler cow herd equal to 30% of the domestic total, 80% of cattle for slaughter and 99% of the imported weaners are concentrated in these northern regions of Italy.

The beef finishing system in the basin of the Po Valley is very different from the extensive beef farming practiced in the dedicated areas of Central Apennines or Southern Italy, based in particular on some traditional Italian white-skinned breeds. The low production of permanent forages and the more reduced areas of the rotation forages determine a greater scattering of these suckler cow farms, which prevents from identifying an extended and consolidated beef system like in the North-Eastern part of Italy (Figure 3). Beef production in these areas consists in suckler cow farms for the production of typical Italian calves that reach the age of slaughtering in the same farm where they were born, or are sold as animals to local fattening farms.

As far as fattening is concerned, three distinct systems can be generally distinguished:

1. The large-scale fattening activity of the Veneto region, where Charolais and crossbreed calves are fattened, mainly imported from France.
2. The medium-sized farms of Piedmont, where fattening of Blonde d'Aquitaine calves imported from France takes place.
3. The small unspecialised farms of Tuscany, where young Chianina bulls are fattened in small numbers. Farm income is here integrated with other agricultural activities.

Three suckler cow systems prevail in Italy:

1. The Piemontese breed suckler cow farms in the Piedmont region.

Table 1. Share of weaners and beef cows among Italian regions in 2001.

Region	Cattle heads > 1 year	%	Imported cattle heads	%	Beef cow heads	%
Veneto	300,000	32	222,000	73	6,000	1
Lombardia	189,000	20	34,000	11	17,000	4
Piemonte	175,000	19	16,000	5	100,000	21
Emilia-Romagna	67,000	7	16,000	5	18,000	4
Northern Italy	751,000	79	298,000	99	146,000	31
Central Italy	86,000	9	2,500	1	82,000	18
Southern Italy	111,000	12	1,500	1	238,000	51
Italy	948,000	100	302,000	100	466,000	100

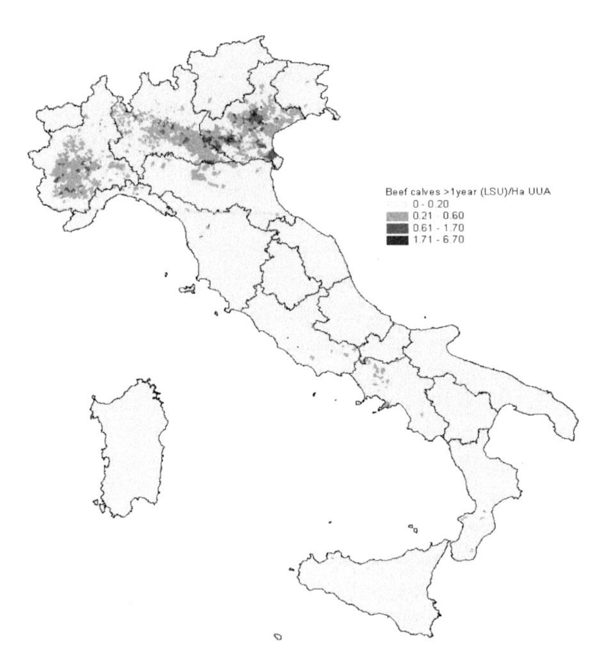

Figure 3. Geographical distribution of beef bulls and heifer cows in 2001.

2. The Chianina, Marchigiana and Romagnola breeds suckler cow farms concentrated in the central regions of Italy (Tuscany, Umbria and Marche).
3. The Podolica breed suckler cow farms scattered all over the South of Italy.

The fattening systems

The systems of the three typologies of fattening farms are completely different. Farm size, feed ration and cattle breeds differ significantly from system to system (Table 2). Obviously, these differences are reflected in the technical performance and the income levels of these farms. In this section, detailed information is provided about the differences between these three fattening systems. The data are based on a yearly monitoring of technical and economic results of beef cattle farms carried out by CRPA on behalf of the Ismea institute of the Ministry of Agriculture (Institute for Agro-industrial Markets of the Ministry of Agriculture).

The largest fattening farms are found in the North-East of Italy, in particular in the Veneto region. Many of these farms rely on hired labour. Another characteristic is the high density of cattle heads per hectare, which makes these farms more vulnerable to the effects of the Nitrate Directive.

Typical family farms are the fattening units specialised in the Blonde d'Aquitaine breed, imported from the South-West of France. Family income is primarily derived from beef production and the cattle density is significantly lower than in Veneto.

Finally, the Tuscan beef farms are typically mixed farm types, where fattening of Chianina calves is a complementary activity to the production of arable crops, olives and tobacco. The fattening units are therefore quite small.

The feed ration in the three fattening systems differs substantially, which is partly due to the beef breed raised on the farm and partly due to the different economic convenience to grow certain crops in the three areas.

Table 2. Farm size and cattle density of fattening farms.

	Veneto	Piedmont	Tuscany
Breed	Charolais Cross breeds	Blonde Aquitaine	Chianina
Calf fattening places (n.)	1,320	380	28
Hectares of forage crops (Ha)	108	65	13
Cattle density per ha (LSUHa forage crops)	7.0	4.5	1.2
Net production (t)	570.5	162.4	8.7

Source: Ismea-CRPA

The predominance of maize silage in the feed ration of the Charolais beef bulls and heifers (almost 7 kg/day) is explained by the high yields this crop can reach in the Veneto region. Maize silage is the basis of the fattening activity in this area and without this crop beef production would not be economically convenient. Another important roughage in the feed ration is beet pulp bought from the sugar industry. The feed ration is complemented with 5,4 kg concentrates a day.
The Piedmont beef fatteners provide a higher energy content in the ration of their Blonde d'Aquitaine beef bulls and heifers with a reduced contribution of roughage and higher quantity of concentrates. Compound feed purchased from the local feed mills has the highest share in the feed ration.
Hay and concentrates are instead the main ingredients of the feed ration of the Chianina calves. Maize silage is scarcely grown in the central regions of Italy, mainly because of a lack of water. Moreover, the code of practice of the local PGI products allows only limited quantities of maize silage in the diet. On some farms, small broad beans (Vicia faba minor) are added to the feed ration.
Obviously, the differences in breeds and feed rations influence the technical performances of the beef fattening activity (Table 3). The highest daily growth rates are registered in the Piedmont farms where a higher energy diet fed to a high quality breed can result in a growth of 1.39 kg per

Table 3. Feed ration of beef bulls and heifers in the three fattening systems.

	Veneto	Piedmont	Tuscany
	Roughage kg/head/day		
Maize silage	6.9	2.2	
Maize mixture	1.1	1.4	
Beet pulp silage	1.9	0.1	
Straw	0.9	0.9	
Hay		1.1	5.6
Total roughage	10.7	5.6	5.6
	Concentrates kg/head/Day		
Maize flour	2.5	1.3	2.9
Barley flour			1.6
Coarse grains	0.4	0.2	0.2
Soybean meal	0.7	0.4	0.6
Proien integration	0.4	0.4	0.1
Compound feed	0.7	3.3	2.1
Dried beet pulp	0.8	0.2	
Total concentrates	5.4	5.7	7.5

day. Another characteristic of this system is that the calves are imported at a weight between 200 and 250 kg, whereas the Charolais calves in Veneto enter in the fattening farm at a weight between 350 and 400 kg.
An interesting aspect of the fattening activity in Veneto, and to some extent also in Piedmont, is that the genetic potential of the Charolais and Blonde d'Aquitaine calves is not fully exploited, as maize silage is the main feed ingredient in the diet (Table 4). If feed was foddered with a significantly higher energy content, these beef bulls and heifers could have growth rates of up to 1.6 or 1.8 kg a day. Under normal circumstances, maize silage is however the cheapest feedstuff, and a feed ration based primarily on cereals would not be economically convenient. This is why the daily growth rates on the fattening farms often do not exceed 1.50 kg a day.
The Chianina calves have more limited growth rates and stay much longer on the farm as calves are purchased at about 250 kg and slaughtered at a live weight of more than 700 kg. The labour productivity of this type of production is rather low, but as the final sale prices are 25-30% higher, it can still be an activity of economic interest. Of course, farm income is integrated with the returns of other agricultural activities on the farm, such as olives, cereals and tobacco. A detailed analysis of production costs and farm incomes is included in chapter 'Impact of the CAP on economic results and location of the production' of this report.

Suckler cow systems

Three suckler cow systems are analysed in this section of the report: the Piemontese suckler cow system, the Chianina cow system and the closed cycle farms in the Marche region where the Marchigiana suckler cow dominates the scene (Table 5). All three farm types are typically family farms with only a limited utilization of hired labour. The largest units are found in Piedmont and small to very small herds are raised in the Marche region.

Table 4. Technical performance of fattening systems.

	Veneto	Piedmont	Tuscany
Calves purchased (n.)	2,145	551	21
Average purchase weight	368	237	257
Fattened bulls sold (n.)	1,988	520	18
Average weight at sale (kg)	641	597	722
Daily growth (kg/day)	1.30	1.39	1.25
Fattening cycle (days)	210	260	372
Mortality rate (%)	1.2	1.7	0.0
Labour productivity (kg/h)	51	35	8,3
Net production (t)	531	162	8.7

Table 5. Farm size and cattle density of the suckler cow farms.

	Piedmont	Umbria	Marche
Breed	Piemontese	Chianina	Marchigiana
Suckler cows (n.)	74	41	11
Calves born (n.)	62	36	9
Forage crops (Ha)	46	38	14
Net production (t)	31.4	19.0	3.5

As far as feed ration is concerned, in Piedmont and Umbria maize silage is an important ingredient of the roughage (Table 6). Hay is mainly fed to the Marchigiana suckler cows. The Piemontese cows rely more heavily on concentrates in comparison with the other two systems, but similar in all three farm types is the use of small broad beans (*Vicia faba minor*), which is an important protein-rich crop, especially in central Italy, traditionally grown for horses but now primarily used for beef cows and calves.
The intercalving period of the Piemontese cows is rather long, which demonstrates problems of fertility. The Piemontese beef bulls are slaughtered at a lower weight than the white-skinned Chianina and Marchigiana bulls. These differences in slaughter weight are reflected at the age at which the bulls are sold to the slaughterhouse (Table 7).

The main evolution: size increase and specialisation of the beef production

Precise statistics on the structure of beef cattle farms are unfortunately not available, as these farms are always integrated in the statistics together with the dairy farms. It is therefore extremely difficult, if not almost impossible, to separate the beef farms from the dairy farms in the official statistics.
An effort has been made here only for the region of Veneto, the most important beef fattening area of the country (Table 8). If we select only the farms having more than 500 cattle heads per farm, we have the majority of beef farms, as only few dairy farms enter into these size categories. Then we notice that the number of farms with more that 1,000 cattle heads is still growing and in the size

Table 6. Feed ration of beef cows of the suckler cow farms.

	Piedmonte	Umbria	Marche
	Roughage kg/head/day		
Maize silage	2,5	3,1	-
Hay	1,8	3,9	7,0
Straw	0,2	1,2	-
Beet pulp	0,2		
Total roughage	4,7	8,2	7,0
	Concentrates kg/head/day		
Maize flour	3,4	2,2	1,0
Barley flour	0,3	2,0	1,3
Soya	0,5	0,6	-
Coarse grains	0,7	0,4	-
Small broad beans	0,2	0,2	0,6
Protein integrator	0,9	0,1	3,0
Total concentrates	6,0	5,5	6,0

Table 7. Technical performance of the suckler cow farms.

	Piemonte	Umbria	Marche
Intercalving period (days)	460	415	438
Age at slaughter of cull cows (years)	7	11	8
Age at first calving (months)	27	27	30
Weight of fattened bulls (kg)	595	653	715
Age of bulls at sale (months)	16	19	20
Net production (t)	31.4	19.0	3.5

Table 8. Farm size evolution in Veneto.

	Farm size		
	From 500 to 999	From 1000 to1999	More than 2000
No. farms			
2000	188	57	19
2005	146	69	28
Number of cattle heads			
2000	124.847	73.229	72.011
2005	96.505	90.561	117.374

Source: Istat

category of more than 2,000 cattle heads the increase in the number of heads is even more pronounced. Farms fattening less than 1,000 heads decline in number and in cattle heads.

The CAP implementation in Italy

As far as the Mid-Term Review application is concerned, the choice has been to adopt a completely decoupled system, starting from January 1st 2005, based on the payments collected by farmers in the 2000-2002 period (single farm payment) and not to resort to the regionalization option (Figures 4 and 5). With regard to the beef sector, the Ministry used the possibility – provided by art. 69 of Reg. 1782/03 – to couple some payments in order to support and improve product quality. The conditions for accessing these additional (coupled) payments are fixed by the Ministry for Agricultural Policies and can be modified each year. Those valid for the year 2005, and confirmed also for the year 2006, mainly aim at indemnifying the suckler cow farms and extensive beef farms (additional premium for beef cows registered in the genealogical books and for beef cows kept in farms that comply with a stock rate < 1.4 LSU/Ha), but finally they are granted also to the intensive fattening farms. Always according to art.69, a slaughter premium is foreseen for male and female bovine animals

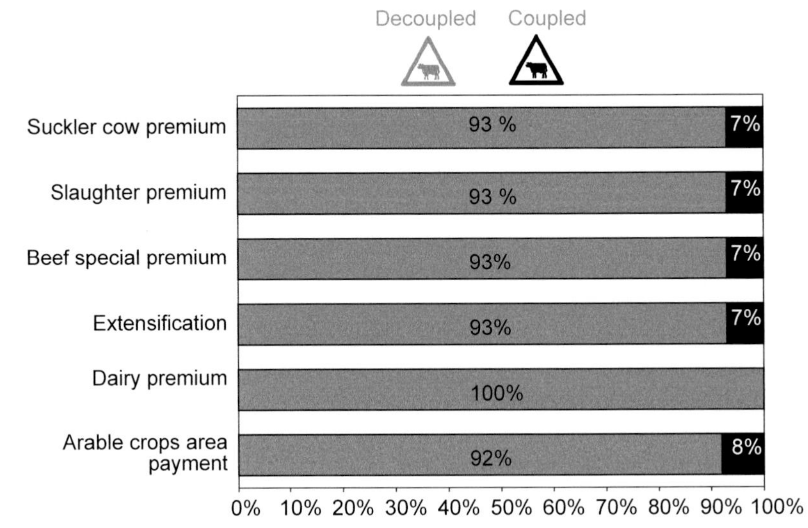

Figure 4. Single Payment Scheme applied in Italy since 2005.

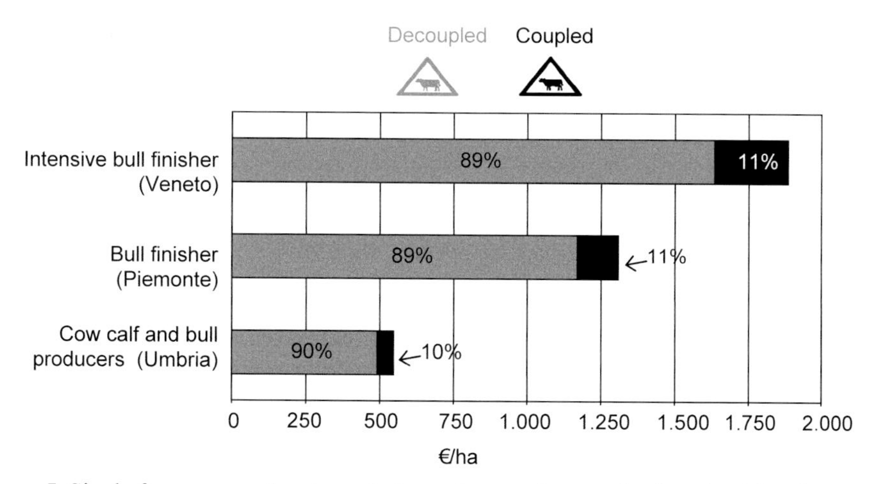

Figure 5. Single farm payment and coupled premium per hectare for three beef farming systems.

slaughtered at an age ranging from 12 to 26 months and complying with optional labelling regulation (ex Reg. 1760/00 EEC). Cattle must be kept in the farm for at least 7 months. The amount of this premium in 2005 was 21 €/head. In total, 7% of the total premium budget has been allocated to the coupled premium ex art.69.

Farmers' and stakeholders' reaction to the implementation of the Mid-Term Review

Farmers' reaction

The first reaction of the beef farmers to the CAP reform was an attitude of 'wait and see'. In the first months of 2005, when decoupling of premia was introduced, many large fattening farms maintained their stables empty in order to verify the new market conditions. Some small farms decided to cease production. This reaction to the decoupling of premia caused a reduction of national beef production equal to 6% in 2005. Imports of weaners for fattening declined in 2005 by 8.5% in total, but in the weight category of over 300 kg imports remained fairly stable. As a consequence, beef imports increased to a new record of 418 thousand tons (+6,6%), as domestic consumption remained stable at 24,5 kg pro capita.
This market shock had a positive effect on the market price of finished bulls, which rose on average by 5%, but for some breeds the increase was up to 11%. At the same time, the price of imported weaners for restalling increased by 7.8%. Due to the reduction of feed prices, the production costs went down by 0.5%, which contributed to the increase of profitability of the beef fattening activity.
After this significant decline of domestic beef production, beef farmers are concerned about the price of imported calves, 90% of which are imported from France. Their price development is a determinant factor for total profitability. A relevant and interesting effect of the premium ex-art.69, which imposes a minimum permanence of 7 months on the farm, is that daily growth rates are lowered in order to receive the premium. This effect is even stronger when imported calves are offered at weights of more than 400 or 450 kg. In 2006 and 2007, the effect of art.69 is slightly mitigated due to its decline from €40-45 down to about €21 per head, but it still cannot be neglected. Its existence does not favour any intensification of production in any case. The predominance of maize silage is

the second factor which hinders further intensification, and both factors contribute to the insufficient exploitation of the genetic potential of imported calves.

Stakeholder position

During the national debate in 2004 concerning the application of the Reform, the positions between the different groups of stakeholders differed significantly.
The national organisations of the agro-industrial cooperatives (ANCA-Legacoop and Fedagri) adopted a common position with respect to the beef sector. In particular, they backed the coupling option of 40% of slaughter premiums of adult bovine animals and of 100% for suckler cow premiums, justifying it as a solution for taking into account the various components of the national beef farming system and preserving the suckler cow system in the less favoured areas, most threatened by the decoupling effects.
Slaughterhouses and the meat industry, represented by ASSOCARNI and FEDERALIMENTARE (Italian Federation of Food Industry), were immediately in favour of the preservation of the slaughter premium coupled with production. The greatest concerns for the slaughtering industry were focused on safety and the continuity of the supplies of cattle for slaughtering, expected by the producers themselves to decline by 10-15%.
The body representing beef butchers and traders (UNICEB) supported the total decoupling proposal of the Italian government. Their arguments underlined the distortion created by the premium system according to Agenda 2000, which, by artificially increasing the demand of French weaners, caused the strong increase in price of the import weaners for fattening farms. The drop in the Italian production would not occur thanks to the improvement of the profitability prospects of the sector due to the expected increase in price and the decrease in the purchase price of French calves. Moreover, the choice of assigning an amount of the national ceiling to a premium coupled with the suckler-cow would have helped in preserving the beef suckler cow herd in the less favoured areas.
Among the different regional breeder organisations, also UNICARVE, a most important association in the Veneto region, officially declared to be in favour of total decoupling. This option was considered the best for the beef finishing farm system (widely represented within UNICARVE), which could have benefited from:

- bureaucratic simplification, thanks to the elimination of all the administrative procedures related to the request of single premiums;
- greater freedom of market choices that were strongly affected by searching direct payments under Agenda 2000;
- greater guarantee of subsidy granting, for what concerned the amounts and the collection time.

Impact of the CAP on economic results and location of the production

Fattening farms

The profitability analysis of beef farms is based on the farms presented in chapter 'The Italian beef farming systems' of this report and carried out by CRPA on behalf of ISMEA (Institute for Agro-industrial Markets of the Ministry of Agriculture) since 2001. For the large beef fattening farms, the sample is formed by a group of farms located primarily in the Veneto region, specialised in fattening of Charolais and crossbreed weaners.
The average size of these farms is about 1,300 fatteners places and ranges from a minimum of approximately 300 to a maximum of more than 2,000 heads. Cattle are mainly French weaners, such as Charolais and crossbreeds, purchased at a weight of 350 kg and fattened for a period of 7/8

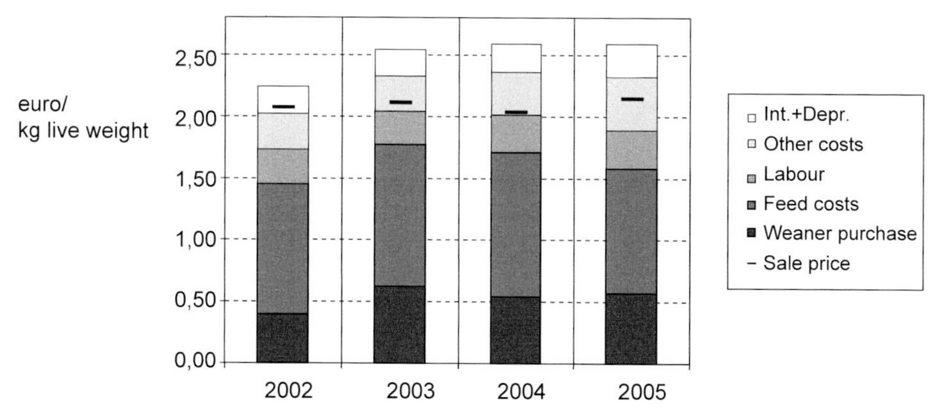

Figure 6. Production cost and return excluding CAP premium (finishing).

months when they reach a final weight of 630 Kg. Average stocking rate on forage crops is around 7 LSU/Ha.

The comparison with the costs of production – expressed per kg of live weight produced – shows that during the period ranging from 2002 to 2004 the sale price of the bullocks did not allow to reach positive profitability margins (Figures 6 and 7).

In the period 2002-2004, profitability decreased due to:

- the overall increase in costs and in particular in the feed and purchase costs of weaners; these are the two factors that most affect the total average cost;
- the contemporaneous fall in the average sale price.

The decrease of the sale price and the increase in costs of production determined the gradual decline of profitability in this period:

- In 2003, the sale price (2.07 €/kg) was equal to 83% of the total cost of production: an amount not sufficient for fully repaying the depreciations and the interests on invested capital, whereas in 2004 (2.03 €/kg) it helped to cover only 80% of the costs: an amount not sufficient for remunerating capital costs and all the family work used in the farm.

In the year 2005, however, the increase of the sale price of finished bulls and a slight decrease of feed prices allowed an overall improvement of profitability of the beef fattening farms. The reduction of domestic production right after the introduction of the decoupling of premia has caused this increase of sale prices, whereas feed prices finally turned to 'normal' levels after the drought of 2003.

Direct payments (coupled before 2004 and decoupled from 2005 onwards) provided by the CMO beef market organisation played an essential role in stabilising the income of the farms. Through the premiums, the profitability margins of the companies remained positive on average. However, between 2002 and 2004, even though the contribution for reducing the costs provided by the CAP passed from 20 to 25%, the net profit of the farms decreased. The unfavourable price performance of finished bulls and the higher purchase price of weaners contributed to the reduction of part of the positive margins given by the premiums. The ratio between the average sale prices and the cost net of the premiums in 2002 was equal to 112% whereas in 2004 it decreased to 105%.

In 2005 the sale price increased considerably (+11%), due the fall of production, and these better market conditions caused an improvement of profitability. The sale price (2.17 €/kg) was equal to 86% of the total cost of production while, considering the single farm payments, total returns covered 113% of total costs.

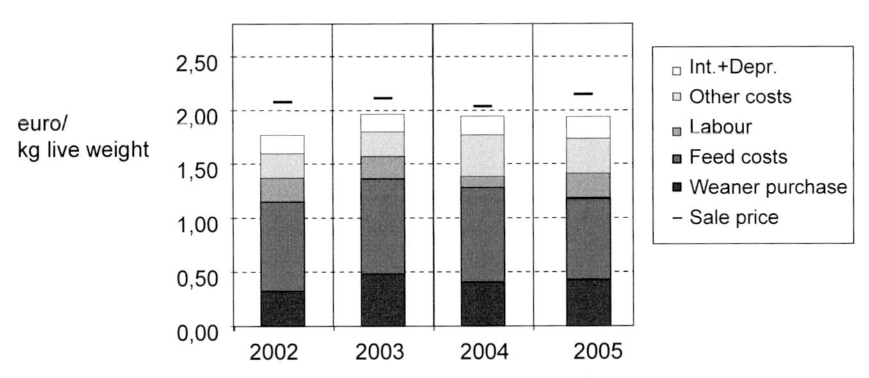

Figure 7. Production cost and return including CAP premium (finishing).

Suckler cow farms

In 2005, the profitability of suckler cow farms increased because of a rise in sale prices and a reduction of production costs (Figures 8 and 9). The sale prices increased by 5.7% for pure Piemontese beef bulls, reaching an average of 3,67 €/kg, whereas the Chianina bulls in Umbria reached a price level of 3,81 €/kg, recording an increase of 0.8%. Overall profitability of these farms improved significantly in 2005 compared to 2004. Both types of cattle breeds are moving into a niche market, as their sale prices are more than 70% higher than that of Charolais or crossbreed cattle imported from France. The interest in Piemontese, Chianina and Marchigiana bulls remains high because of their highly appreciated cuts in the Piedmont and Tuscan cuisine.

Figure 8 reveals that the price alone is not sufficient to cover the very high production costs per kg beef, which varies between 4,69 €/kg and 5,13 €/kg for Piemontese beef and 4,89 €/kg and 5,61 €/kg for Chianina beef. In 2005 production costs declined by 8.5% in Piedmont and by 12.8% in Tuscany, mainly because of significantly lower roughage and concentrates costs.

The CAP premiums are crucial for the survival of these suckler cow systems. Figure 9 shows how these premia are able to improve profitability almost to the point where costs and returns equal, in particular in the year 2005. The decoupled premia are the sum of premia for suckler cows, slaughtered

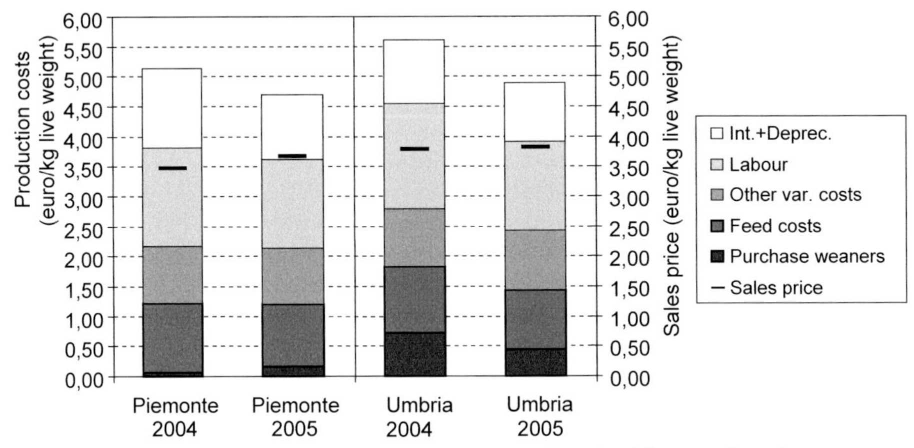

Figure 8. Production cost and return excluding CAP premium (suckler cow farms).

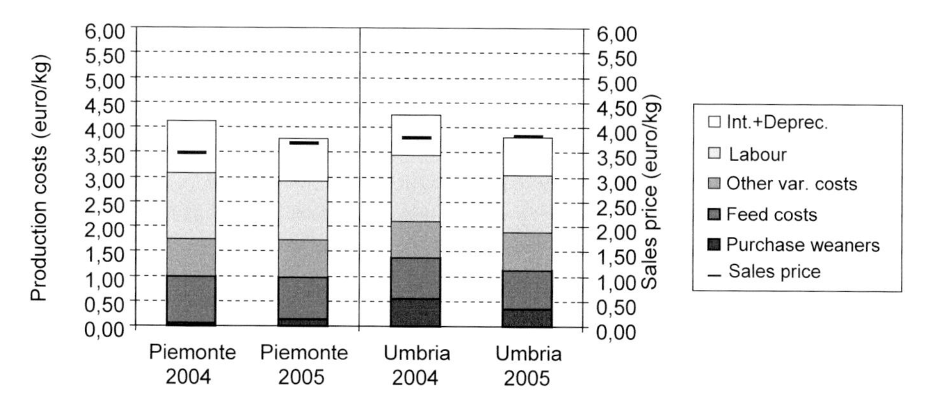

Figure 9. Production cost and return including CAP premium (suckler cow farms).

bulls and heifers, live bulls, article 69, extensification, premia for maize silage and in some farms also for tobacco.

Future of fattening activity in Italy

Most of the beef fattening farms are highly specialised in beef production. The utilised agricultural area of these farms is dedicated by more than 90% to feed production, either roughage (maize silage) or concentrates (cereals). This strategy is pursued in order to reach the highest independence possible of the feed market. For this reason, the decoupling of premia does not open a strategy of diversification, as the high rate of specialisation remains the main objective of fattening farms. Feed and weaner prices determine the major part of profitability of the beef farms. The future of beef fattening heavily depends on the development of these prices.
The effective application of the Nitrate Directive will take place in 2007 and 2008. The Nitrate Vulnerable Zones (NVZ) have been extended substantially in Veneto, Lombardia and Piedmont, as these regions in the past did not apply this Directive properly. Many fattening farms in Veneto are located in the NVZs, where a maximum of 170 kg N of manure can be spread. Manure storage capacity has to be extended to 120 days. The transport or treatment of excess manure and the investment in extra manure storage will increase production costs for fattening farms located in NVZs. The individual farm situation will heavily depend on the cattle density per hectare. The large farms will continue to benefit from conspicuous Single Farm Payments which will enable them to face the Nitrate Directive problem with less financial efforts than the small farms. However, eventual decisions about maximum ceilings of CAP payments might most affect these farms. It is expected that modulation of payments generate more limited reductions in subsidies. The second reform of the CAP to be carried out in 2008 will be of great importance.
Some small farms will cease production as not all of these farms will be able to sustain the increase of costs related to the Nitrate Directive. A further size increase in order to exploit economies of scale will be blocked by the effect of the Nitrate Directive.
Beef consumption in Italy will probably decline slightly due to the strong competition of lower priced poultry and pig meat. Meat imports from South America will increase, in particular when a further liberalisation of trade will be decided within the Doha round. Actually this meat is marketed in the restaurant and catering circuits, but in the next years this beef will appear on the supermarket shelves as well.

The final result of these developments will consist in the beef fattening activity in Italy continuing at a lower level than the actual one, as the sector has to absorb the extra costs related to the application of the Nitrate Directive and also face a more fierce competition from abroad. A further intensification strategy is not foreseen as the maize silage system will remain the most convenient way of feeding.
The suckler cow system in Italy will continue to feed a niche market of high priced quality beef, based on the local beef breeds of Central Italy. Expansion of this system is not probable, as the natural and climatic conditions pose limits to its growth. For the same reason, the domestic production of weaners of these breeds will not grow significantly, as Italy does not have a comparative cost advantage in this production. Scarce rainfall in the summer, low roughage production, and a highly fragmented farm structure are not the ideal conditions for a profitable production of weaners able to compete with weaners imported from abroad.

References

De Roest, K., C. Montanari and E. Corradini, 2007. Costi di produzione e di macellazione del vitellone da carne – Opuscolo CRPA n.2.49 , Tecnograf, Reggio Emilia.

ISMEA – Osservatorio Latte, 2006. Il mercato della carne bovina, Rapporto 2006, Franco Angeli, Milano.

Impact of the new CAP on Spanish beef farming systems

Subdirección General de Pagos Directos, Vacuno y Ovino, Dirección General de Ganadería

MAPA (Spain), sgpdvo@mapya.es, www.mapa.es

Abstract

The report gives a general picture about the current situation of the Spanish beef sector, and the impact of the last CAP reform on it. Starting with the presentation of the main figures of the Spanish beef sector, the report describes the geographical distribution of Spanish beef farming systems, with two separate and complementary sub-sectors: the cow-calf producers and the beef fatteners. The report summarises the main decisions adopted in the last CAP reform that started to be implemented in 2006. Among the various options contemplated by the community rules in the beef sector, Spanish farmers decided to choose the maximum coupling possible for suckler cow premiums in order to avoid the risk of abandonment of the agricultural activity, particularly in less productive regions. The implementation of article 69 of the CAP was also adopted, for specific activities that are important for the protection of the environment, and to improve the quality and marketing. The report summarises the main important tasks for the future under the new scenario, such as the maintenance of national suckler cow census, the improvement of cows' fertility, and the increase of the profits of fattening weaners.

The Spanish beef production

The Spanish beef sector, as can be seen in Figure 1, occupies the second place behind the pig sector within the livestock sectors for its contribution to the final livestock production, representing in 2006 19% of the Final Livestock Production and 7% of the Final Agricultural Production.
In the last decade, the number of cattle in Spain increased continuously (Figure 2), in spite of small declines in certain years. There are now 6.4 million heads. In particular, the number of suckler cows

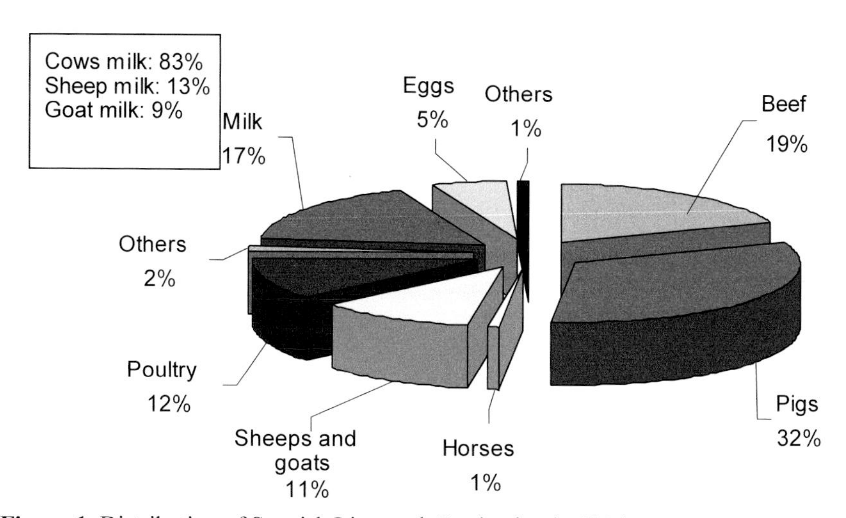

Figure 1. Distribution of Spanish Livestock Production in 2006.

increased greatly, reaching 2 million heads (half a million over the quota of premium rights for suckler cows). In EU-25, the Spanish beef sector is fifth as regards the number of heads and fourth as regards meat production, behind countries with an old beef tradition such as France and Germany.
Compared to 2005, in 2006 Spanish beef production fell from 2.766.100 to 2.576.000 heads (Figure 3). This drop represents a decrease of more than 8% in tons.
Spain exports 20% of its beef production (2006). The main trading partners are countries of the EU (94% of the exports are directed to the intra-Community market, mainly countries around Spain such as Portugal, France and Italy). The remaining 6% is sent to third countries, such as Russia. The latter part has dropped significantly in recent years as a result of quotas and the reduction in export refunds. Figure 4 shows the main destinations of Spanish beef exports in 2006.

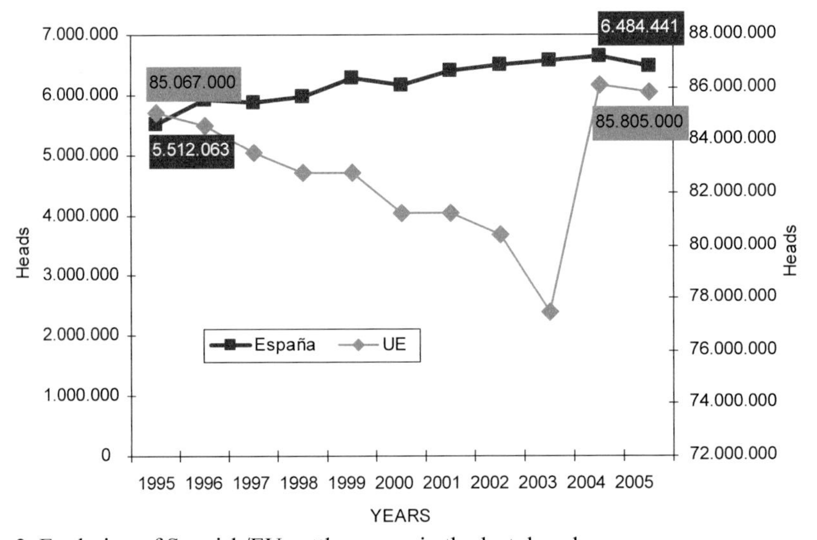

Figure 2. Evolution of Spanish/EU cattle census in the last decade.

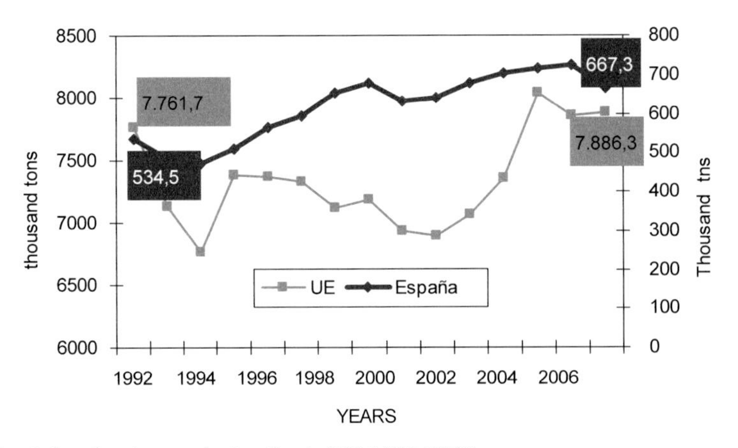

Figure 3. Beef slaughtering evolution Spain/EU 1992-2006.

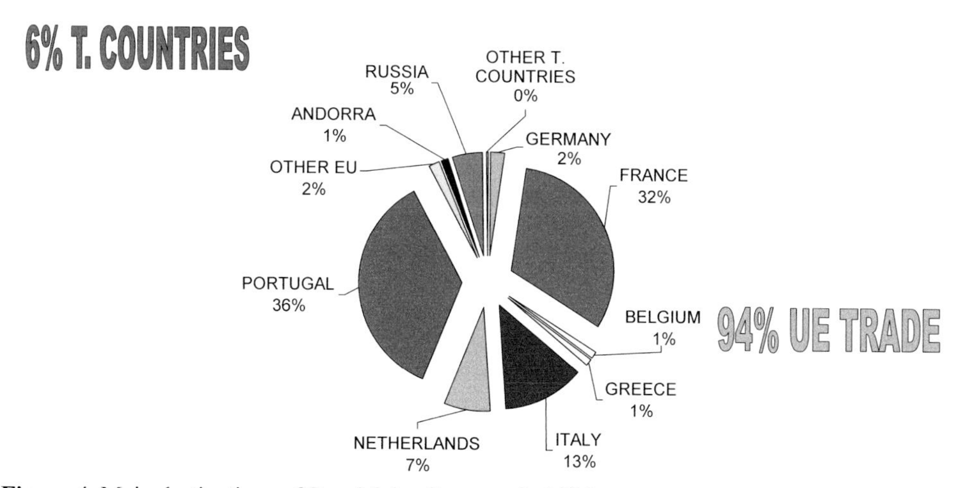

Figure 4. Main destinations of Spanish beef exports in 2006.

The Spanish beef farming systems

Two separated but complementary sectors with specific characteristics must be distinguished (Figure 5):

The cow-calf sub-sector

The cow-calf sub-sector is principally located in the mountainous areas of the northern peninsula and in the meadows of South-west ('dehesas'). These are farms with a wide territorial basis. Feeding is based on crops produced in the farm, mostly supplemented with concentrates in certain months of the year when there is lack of pasture or in case of increased needs of the flock (gestation).
Most of the animals are indigenous breeds, helping to maintain the Spanish genetic heritage and promoting environmental conservation by using natural resources. In addition, production contributes

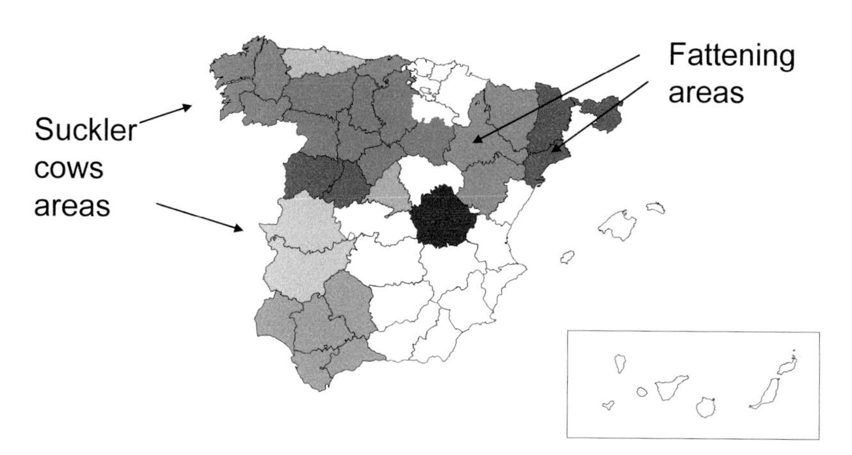

Figure 5. Geographical distribution of the Spanish beef sector.

effectively to the organization and management of the territory, especially in disadvantaged areas with few alternatives.
The main production on these farms consists in weaners that stay with the cows until the age of 5-6 months, when they are sold to beef farms where they are fattened and finished.
Three areas can be distinguished:

- Cornisa cantábrica (number 1 and 2 in Figure 6):

In the western area of the Cantabrian Cornice (Galicia), farm size ranges from 120 to 200 cows, with a rather high stocking rate (0.5 hectares per cow). Because of regular rainfalls, feeding needs can be covered by pastures all along the year with little supplementation.
The most common breed is the pure 'Rubia Gallega' or Rubia Gallega crossed with other breeds to improve the quality of meat. Calves remain with the cows until slaughter (at the age of 8-10 months). Since the age of 6 months they are fed with concentrates. The main production is 'Veal of Galicia' (Protected Geographical Indication).
Areas in the eastern part of the Cornice are the poorest ones (mountainous areas). Communal pastures are frequent and herds are occasionally sent to the mountains during the summer season. The farms are smaller (15 to 10 animals) and stocking rates are lower (1 hectare per cow on average). Supplementation is necessary for several months throughout the year. The main breeds are 'Asturiana', 'Pardo Alpina' and 'Tudanca', as well as industrial crossbreeds.

- Castilla y León (number 3 in Figure 6):

The size of the farms varies from 120 to 200 cows per farm. These areas are less productive because of the climate, with stocking rates not higher than 2 cows per hectare. The production of pasture is more seasonal, resulting in long periods of supplementation throughout the year.
The most frequent breeds are the 'Avileña-Negra-Ibérica' and 'Morucha', as well as crossbreeds (Charolais and Limousine bulls are used to improve the quantity of meat).

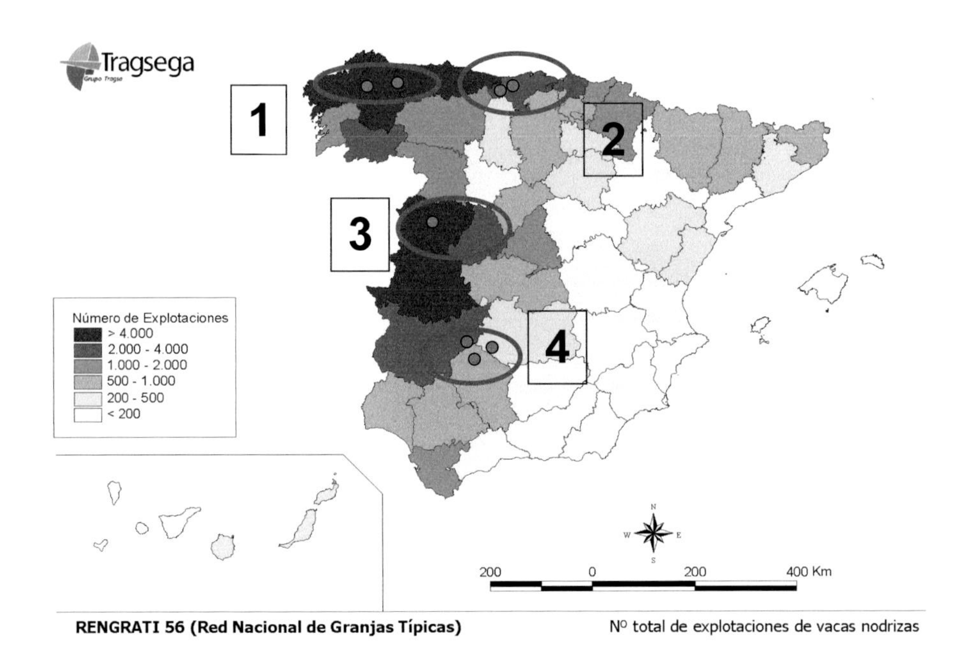

Figure 6. Geographical distribution of cow-calf farms belonging to the National Network of Typical Farms (RENGRATI).

Mixed farming systems are common in these areas, such as calves born on the farm kept to be fattened with calves bought outside the farm on local markets.

- Andalucía, Castilla La Mancha y Extremadura (number 4 Figure 6):

The average size of these farms is 70-100 cows and the stocking rates are around 1.5 hectares per cow. Those typical Mediterranean areas are called 'Dehesas' and the cows are managed together with sheep and Iberian pigs. Due to lack of rain, these areas suffer from drought. Cows and calves need to be fed with purchased feeds, which represent a high cost for production, compromising in some cases the viability of the farms.

The main cow breeds are pure 'Retinta' and Retinta crossed with Charolais and Limousine bulls. In recent years, some cooperatives in these areas have developed communal fattening centres in an attempt to improve their economic returns.

The beef finishing sub-sector

The Spanish beef finishing sub-sector is more specialized and competitive, with an important capacity of exportation and continuous growth in the last years.

The farms are concentrated near the large areas of consumption, where the main slaughterhouses are also concentrated, in the Autonomous Regions of Cataluña, Aragón, Castilla y León, Madrid and Galicia. Given the special climatic conditions of Spain, animal feeds are only concentrates based on cereal components.

An important part of the calves finished on these farms come from other EU countries. For example, in 2006 the Spanish beef finishing producers bought around 800,000 calves from Poland, France, Germany, Ireland and Italy.

Two different areas can be distinguished in the beef finishing sub-sector:

- Cataluña y Aragón (number 1 in Figure 7):

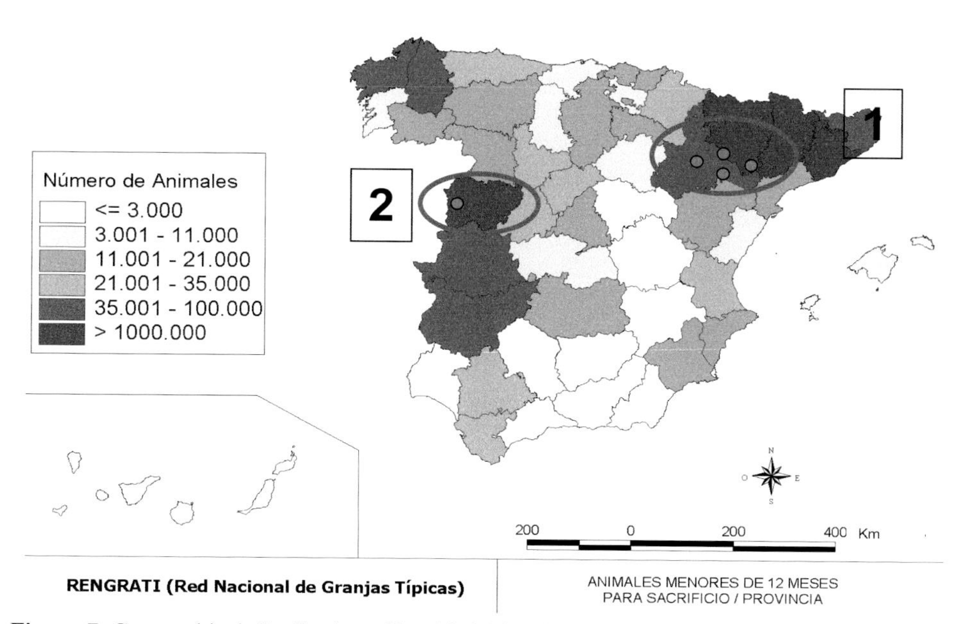

Figure 7. Geographical distribution of beef finishing farms belonging to the National Network of Typical Farms (RENGRATI).

The size of these farms is highly variable. They range from 500 to 1000 animals. There are also feedlots working in integration systems that feed around 7.000 animals per year in a set of farms in the area.
The calves fed on those holdings come from the 'pintos' markets or are 'milked crossed' calves coming from milk holdings in the Cantabrian cornice in Spain. But they mainly come from other EU countries (Simmental or Blonde d´Aquitaine from Poland, Ireland, Germany or Italy).
The number of males or females in a year depends on the purchase prices and the evolution of markets.

- Central Area (Castilla y León, Segovia y Toledo) (number 2 in Figure 7)

The size of these holdings varies between 200 and 800 weaners per farm and year. They fatten weaners coming from the own farm or bought on local markets. Feeding is based on concentrates. Some communal fattening centres linked to PGI production are starting to develop.

The main evolution: size increase and specialization of beef production

The Spanish beef sector has known a remarkable growth in the last years. The sector's capacity to adapt to the challenges imposed by the increasing opening of markets and the strong competition from abroad, is attributed to technological and structural improvements that allowed the sector to become more specialized and competitive.
In Spain, like in the rest of Europe, dairy farms have a great influence on meat production. Most of the farms are located in the Cantabrian Cornice. In addition to the cull cows they produce, they also sell the calves at the age of one month to be fattened in finishing farms.
Due to the increase of milk productivity, the number of dairy cows has decreased dramatically from 1.4 millions in 1992 to less than 1 million today (Figure 8). In the same time, the number of suckler cows increased from 1.2 million to nearly 2 millions, as seen in Figure 8.
The increase in the number of suckler cows in the recent years was not only clearly supported by the CAP implementation, but also helped by the market and the competitiveness of the sector. At present there are half a million suckler cows over the national ceiling of aids.

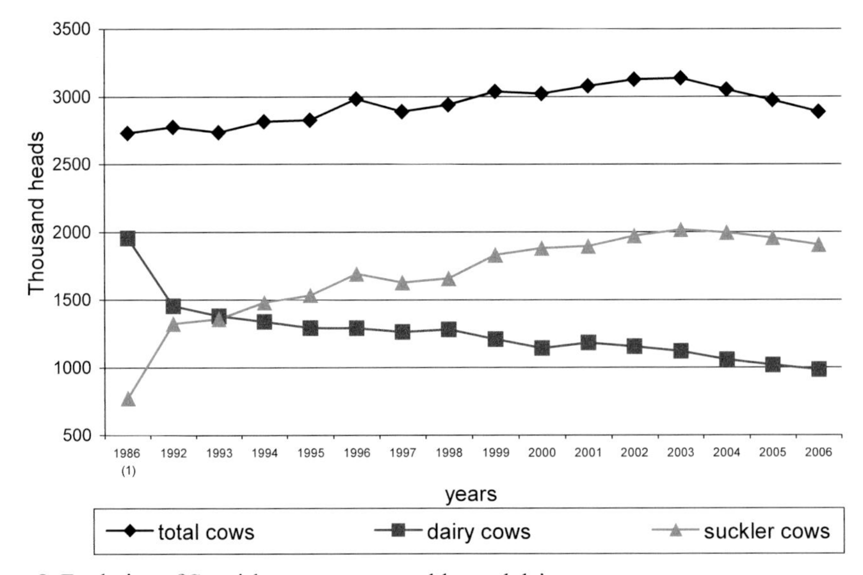

Figure 8. Evolution of Spanish cow census: suckler and dairy cows.

As regards the type of animals produced in the Spanish beef market, the tastes and habits of Spanish consumers differ from those in other EU countries. In Spain, bulls and heifers are slaughtered at younger ages than in the rest of the EU (12-14 months). The average weight of the carcasses is 260 kg compared to 283 kg in Europe. But there is no tradition of consuming 'white' calves as in other countries like France, Belgium or Italy. Another particularity is the high proportion of heifers (31% compared to 16% in the rest of Europe) and the absence of castrated animals. In Figure 9, differences can be seen between Spain and EU in slaughtering distribution.
It must be noted that the cycles of beef production in Spain are shorter than in the EU average (9 months compared with 18 on average in the EU), thus helping farmers to adapt more quickly to changes in market regulations.
The evolution of the beef sector in the recent years is linked to a growing entry of calves from other EU countries, mainly from France, Poland, Germany, Italy and Ireland.
However, live animal imports decreased by 30% between 2005 and 2006, due to the decrease of old imported animals (Figure 10). In this way, at the same time the numbers of animals less than 80 kg still increased by 3% and the numbers of animals 80 kg-160 kg rose by 0.5%.

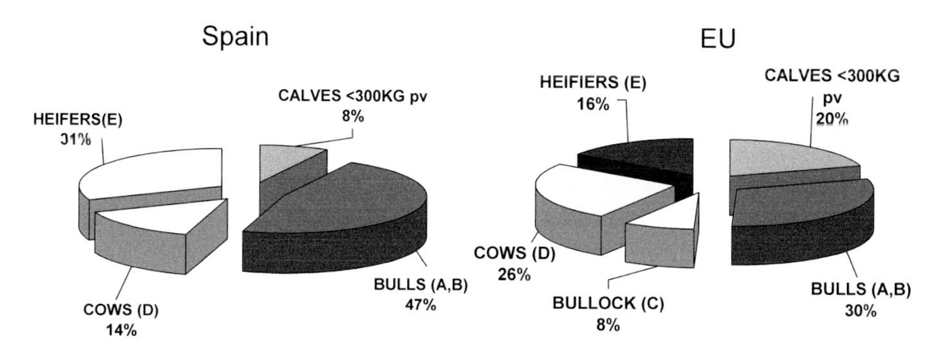

Figure 9. Slaughtering distribution by ages Spain/EU 2006.

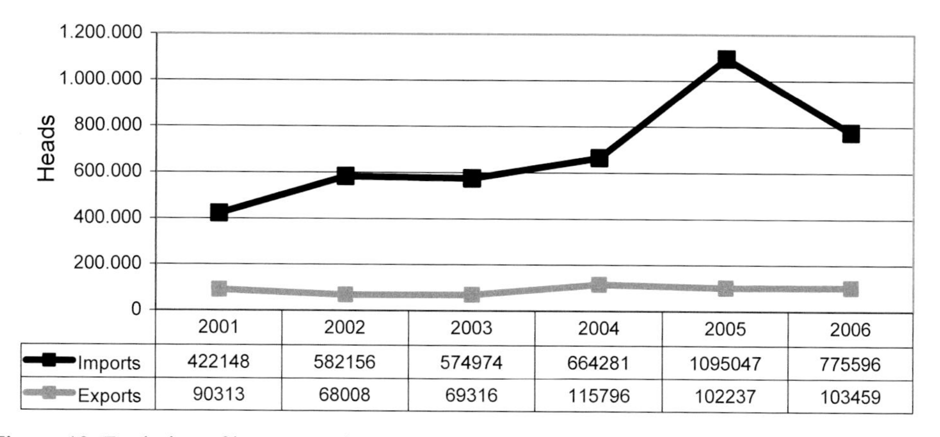

	2001	2002	2003	2004	2005	2006
Imports	422148	582156	574974	664281	1095047	775596
Exports	90313	68008	69316	115796	102237	103459

Figure 10. Evolution of imports and exports of cattle in Spain 2001-2006.

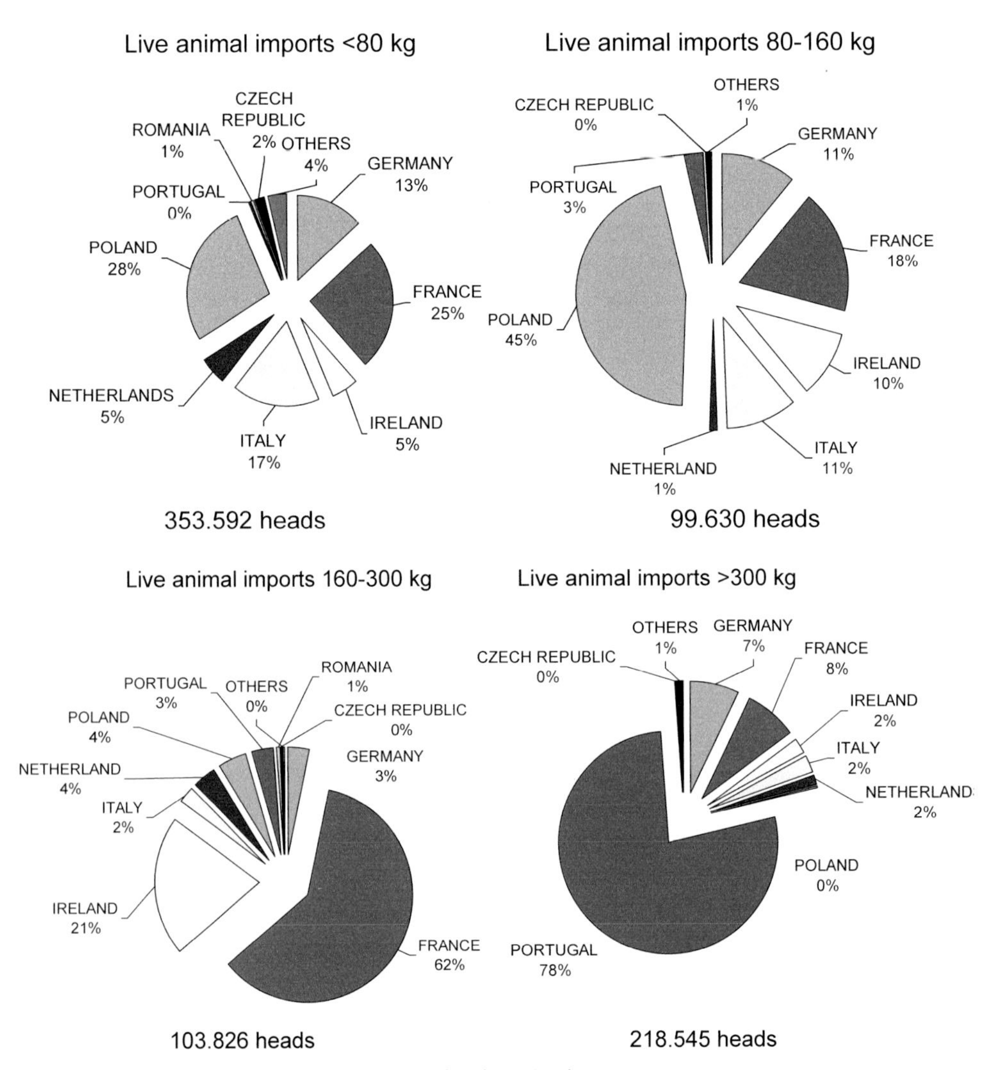

Figure 11. Community imports of young animals to Spain.

The CAP implementation in Spain

In 2006, the CAP Reform was implemented for the first year in Spain, concerning mainly the livestock production.

Spain tried to find a model of implementation that satisfied the general interest: maintaining the agricultural activity in the medium and long term in the entire territory and allowing the optimisation of community resources. Direct aids have had a significant influence on developments in the beef sector in recent years. The volume of FEOGA-Guarantee funds to the European beef sector has been increased in successive reforms. For this reason, the last reform of the CAP in 2003 undoubtedly represents one of the greatest challenges for the Spanish beef sector to adapt to new constraints. Main decisions adopted are:

- implementation in 2006;
- model: historic and on a national scale (Canary islands excluded);
- maximum coupling possible;
- implementation of article 69.

Among the options contemplated by the Community rules in the beef sector, Spain maintained coupled certain aids:

- up to 100% of the slaughter premium for calves;
- up to 100% of the suckler cow premium;
- up to 40% of the slaughter premium for cattle adults.

Therefore, the special premium for male cattle, the extensification premium and additional payments (envelopes) that were subject to certain criteria in each Autonomous Community were converted into the Single Farm Payment (decoupled payments).
The main reasons for this choice were the following:
With a system of total decoupling aids, there was a high risk of abandonment of agricultural activity, particularly in the less productive regions. Therefore, Spain chose to maintain the aids coupled as much as possible, in order to favour the conservation of natural environment and to help maintain populations in those rural areas. Nevertheless, the new rules represent an unprecedented change of deregulation in agricultural activity.

Coupled suckler cow premiums

Suckler cow farms have an important territorial basis. They are mainly located in disadvantaged or mountainous areas and therefore contribute effectively to maintaining population in the countryside. They also contribute to the conservation of permanent pastures in those areas.
Moreover, it is necessary to sustain the activity of cow-calf herds because most of them are of indigenous Spanish breeds, and it is important to preserve the genetic heritage of such breeds. Lastly, maintaining the suckler cow farms in these regions helps to provide the fattening enterprises with Spanish calves, thus limiting the dependence on foreign trade.

Coupled slaughter premium for cattle adults

Maintaining the slaughter premium coupled allows to support the entire beef sector, including the dairy sector. In addition, this premium helps meet the requirements set by the system of identification and registration of cattle, which directly contributes to health monitoring.

Implementation of article 69 of the CAP reform in the Spanish beef sector

The implementation of Article 69 is based on two pillars:

- The possibility to retain up to 10% of the sectorial components of aids. Spain decided to apply a sectorial retention of 7% to all payments of the beef sector (approximately €54 million).
- This amount of retention payments is dedicated to specific agricultural activities: protection or improvement of the environment, improvement of the quality and marketing of agricultural products. Two measures have been adopted for the Spanish beef sector:
  a. Support for the extensive holdings:
     47 million euro will be spent to encourage the maintenance of livestock activities involving a benefit from an environmental point of view.
     The additional payments are linked to suckler cow holdings, with or without premium rights, and concern the farms with a maximum of 1.5 LU/ha. The amount per cow is modulated in

proportion of the stocking rate of the farms, with a maximum of 100 premiums per farmer. The aim of that payment is to promote a rational use of land.
In 2006, 1.684.723 Spanish suckler cows have received this additional payment. The amounts per cow were as follows:
- for the first 40 heads of the flock: 32,62 euro/cow;
- from 41 to 70 heads of the flock: 21,85 euro/cow;
- from 71 to 100 heads of the flock: 10,76 euro/cow.

b. Support for the production of beef of differentiated quality:
The overall amount for this measure will be 7 million euro. The aim is to promote the improvement of quality and marketing of beef, by supporting the production of quality beef. This primarily concerns fattening systems that have adopted specific quality regulations as:
- protected designations of origin or protected geographical indications;
- organic production;
- voluntary labelling requirements involving some higher standards than those required in the general rules.

All those systems provide additional guarantees to consumers regarding the origin of the meat, the method of production, the feeds used, the management of the system and the traceability throughout the chain. These systems guarantee not only a differentiated quality product, but also an improved marketing.
A maximum of 200 animals per farm is considered, except in community fattening centres where this limit is fixed for each partner.
In 2006, 475.000 slaughtered cattle involved in a differentiated-quality program have received this payment at an amount of 14,73 euros per head. This number represents 19% of the total Spanish beef production.

The impact of the CAP on economic results and location of production. Consequences for each system.

At the point of deciding decoupling the special premium for male cattle and maintaining the suckler-cow premium coupled, there could be tensions in the productive chain between suckler cow producers and fatteners. It was therefore feared that the prices of calves could decrease.
As can be seen in Figure 12, the price of weaners was stable during 2006; 10% higher than in 2005. On the other hand, the prices of beef meat, for example carcasses of AR3 cattle, were 11,5% higher than in 2005. Thus, despite of the decoupling of premiums for fattening, the high price of beef maintained the interest of producers to fat animals and this was the reason of the higher prices of weaners. The average price of concentrates, which is one of the most important input factors for producers for the near future, remained quite stable until the beginning of 2007.

Consequences for cow-calf producers

The main important objectives of the Spanish cow-calf sector in the near future are the following:
- Try to maintain the national cow census that ensures the supply of a significant number of calves per year for fatteners, and contribute effectively to promote environmental issues and maintain population in the countryside.
- Make efforts to improve the fertility of suckler cows and thus increase the availability of endogenous calves produced. The factor of supplementation in cows is important, especially in relation to the agro-climatic conditions in Spain (in particular the severe drought that occurred in the last years). Such drought could become a structural factor having a significant impact on livestock production, mainly a notable increase of production costs.

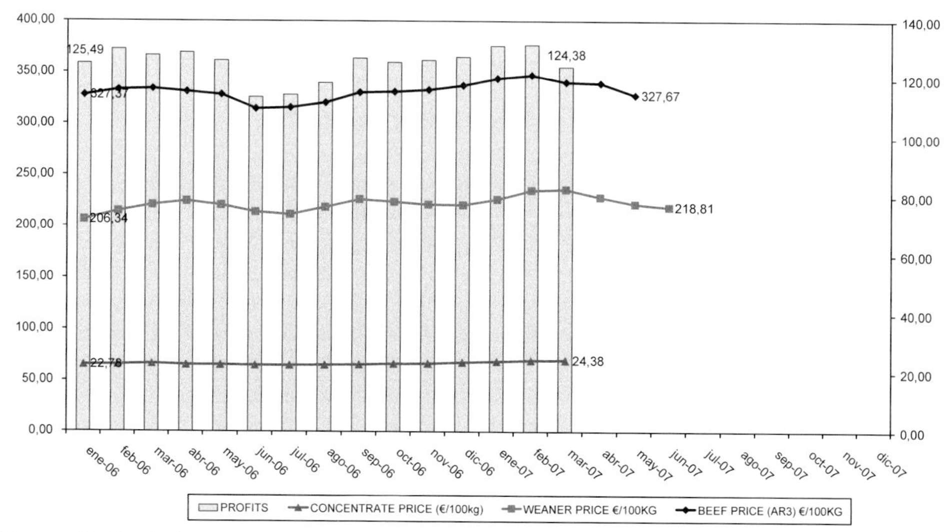

Figure 12. Evolution of prices for concentrates, weaners and beef (AR3 carcasses). Prices since January 2006 in Spain.

- Maintain the level of premiums, while in many suckler cow farms the profit consists in the premium.
- Try to complete the fattening of weaners through community fattening centres, or PGI production linked to indigenous breeds highly appreciated by consumers.

Consequences for fatteners

Limiting factors for the Spanish beef fattening sector in the coming years would be the following:

- Probable difficulties in the supply of calves at EU level:
  - Health problems related to the appearance of the bluetongue disease in Central Europe could limit the normal trade of calves.
  - Increasingly demanding conditions in the regulations concerning animal transport and welfare, resulting in higher costs.
  - The potential growth of some of the main suppliers of calves, primarily in new EU member states, including Poland, who could try to position themselves as major producers by fattening calves born in their farms instead of exporting them to Spain.
  - The search in third countries for new markets of calves for fattening can also be a possibility for the coming years, but health problems and difficulties in transport can be an important limiting factor.
- Feeding is also going to be one of the major constraints in the coming years, because of the special agro-climatic conditions in Spain. In particular, a large part of Spanish beef production is based on cereal concentrates and could enter in competition with bio-energy production in the near future.
- A possible solution for the future could be to seek a balance between the costs of inputs, calves and concentrates, exploring models based on economies of scale. Integration systems have become more and more important in Spain in the last years. As intensive systems are highly dependant on the purchase price of animals and concentrates, they apply models that improve the management of purchases in order to optimise results. In the following years it will be important to analyse

integration models in order to reach useful conclusions. Another important issue is that these systems must be linked to the chain. Vertical integration in the sector could be a solution.

Conclusion and future of Spanish beef production

There is no doubt that, in the context of the latest reform of the CAP, the single farm payment and the cross-compliance requirements are going to be the most important factors for the future evolution of the Spanish beef sector. Other significant factors include the following:

- Smaller farms, and therefore less specialised and competitive, will probably disappear. On the contrary, business concentration will result in modern management and quality systems.
- Trade exchanges are increasingly important, therefore meat imports from third countries are of high importance. 25% of Spanish beef imports come from third countries, mainly from South America: Brazil, Uruguay and Argentina (in this order) represent an important part in high quality beef supplies. Given the intensity of foreign competition, the Spanish beef sector has a number of advantages coming from the EU production model and regarding food security, quality, respect for the environment and animal welfare. These elements add value to the Spanish beef production and differentiate it from the beef production of third countries.

In conclusion, the Spanish beef sector has demonstrated a good capacity to adapt to past CAP reforms. It is therefore legitimate to believe that it will adapt well also to the last reform, through a gradual but necessary restructuring.

Effects of the CAP Reform on Swedish Beef Production

T. Strand[1] *and P. Salevid*[2]

[1]*Taurus Köttrådgivning AB, C/O LRF Sydost, Box 974, 391 24 Kalmar, Sweden*
[2]*LRF Konsult, Mässhantverksgatan 5, 598 40 Vimmerby, Sweden*

Abstract

The CAP reform has contributed to the decrease of beef cattle in Sweden. Just before and during the implementation of the reform, the will to make capital investments was low, owing to a great uncertainty as to the future conditions for beef production. The will to invest has now returned, and many beef producers have attained an improved profitability after the reform. Swedish beef production is presently encountering several threats, one being the shortage of calves! More calves need to be born in Sweden in order to secure a solid and competitive production. However, we do believe that the number of suckler cows will still increase as a result of the continued focus on grazing land in Sweden. The future focus on bioenergy will mainly affect the prices of feedstuffs, but to some extent also the prices of land. An estimation of future costs in Swedish beef production is therefore uncertain.

Swedish beef production

In terms of surface, Sweden is one of the biggest countries in Europe. Forests take up almost half of the country, and more than one third of the total area consists of mountains, lakes and wetlands. Approximately 3 million hectares are agricultural land. The type of farming varies greatly between the northern and the southern part of the country, primarily due to considerable differences in climate. The annual mean temperature in northern Sweden is minus 2.5 °C, while in the southern parts of the country it is approximately 8 to 9 degrees higher.

Only 1.5% of the area of Sweden, that is about 500,000 hectares, consists of permanent grassland (Figure 1). Natural grasslands have a very high biological diversity with many red-listed species. Cultural traces in the landscape are abundant and highly appreciated, as are also the possibilities of recreation. There is a strong will among politicians, authorities, as well as the general public, to preserve these valuable biotopes. To preserve the permanent grassland is the only quantified political goal in Swedish agriculture. This goal played a major role in the application of the Mid-Term Review reform in Sweden.

Sweden has a population of 9 million people, of which approximately two percent are economically active in agriculture. There are somewhat more than 70,000 farmers, and the majority of farms are family-owned. Agriculture and forestry are often combined, which is a typical situation for Swedish beef production.

In 2006, the production of beef including veal amounted to 138,000 tons, while the consumption of beef was 228,000 tons; approximately 25 kg per capita (Table 1). Of all beef consumed, slightly more than 55% was Swedish beef. Low self-sufficiency makes the Swedish market a dumping ground for beef exporting countries. As a consequence, together with a marked concentration of the commerce, it is difficult to keep the Swedish producer prices up. Most imports to Sweden come from Ireland, Denmark and Germany (2005) and, in relation to this, it is important to emphasize the fact that Sweden is presently not capable of producing enough beef for its home market.

There are roughly 1.6 million head of cattle in Sweden, divided among approximately 25,000 enterprises (Table 2).

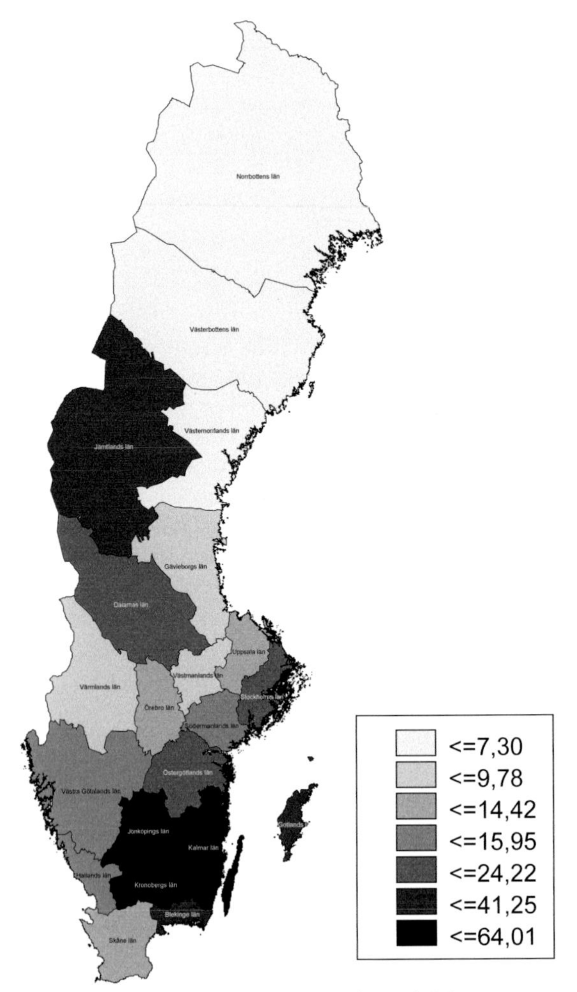

Figure 1. Grassland in Agricultural Land (%).

Table 1. Beef Market in Sweden (thousand tons).

	2004	2005	2006
Production	143	137	138
Import	95	105	102
Export	11	12	11
Consumption	228	231	228

There are roughly 565 000 cows in Sweden. Numerically, the largest breeds in Swedish dairy production are Swedish Holstein and Swedish Red Cattle. The approximate distribution between these breeds is 50/50. About 70% of all cows in Sweden are dairy breeds and 30% are beef breeds. The most numerous beef breed in Sweden is Charolais, followed by Hereford, Simmental, Limousin, Angus, Highland Cattle and Blonde d'Aquitaine, and some breeds with a very small population.

Table 2. The Swedish Systems for Breeding of Beef Cattle.

	Number of farms	Percentage
Suckler calf producers	12,447	49%
Dairy producers	8,027	32%
Beef finishers	4,608	18%
Total	25,000	100%
of which combined calf producers and beef finishers	19,092	76%

Only about 20,000 beef cows are purebreds, and the remaining part consists of crossbreeds used in the production herds.

Swedish beef production is closely integrated with dairy production. Approximately 70% of all slaughter in Sweden comes from dairy production, either as young animals of dairy breeds or cows culled from dairy production.

The average herd size of a Swedish suckler cow herd and a dairy herd is 14.3 cows and 48.3 cows, respectively. The average breeder of young cattle delivers somewhat more than 22 animals for slaughter each year. Table 2 shows that most suckler calf producers and dairy producers also deliver young stock for slaughter. 76% of Swedish beef farms combine the production of calves and young stock. This means that 93% of all dairy herds and suckler cow herds also deliver young cattle for slaughter.

Of all slaughtered animals in Sweden, 42% are young bulls, 11% are bulls, 10% are heifers, and the rest are cows (Figure 2).

The production of suckler calves is mainly carried out on small farms, with suckler calf production not being the main occupation. Depending on the geographical location in Sweden (Figure 3), the indoor period varies between five and seven months. The feeding regime consists of silage, hay and/or straw, together with grain and minerals. The majority of beef calves are born between January and May and are weaned during the autumn, with a weaning weight varying between 200 and 400 kg depending on sex, age, breed etc. During the winter, the suckler cows are either kept in deep-straw bedding systems, in free stalls or tied up on stall-floor.

Dairy production is mainly carried out on family-owned farms. Calves are born with a relatively even distribution during the year, but with a minor peak in early autumn. The calves are either sold live at a weight of 70-130 kg through livestock dealers or in accordance with agreements between farms to specialised breeders of young cattle; otherwise the calf is kept and bred for slaughter in its birth herd.

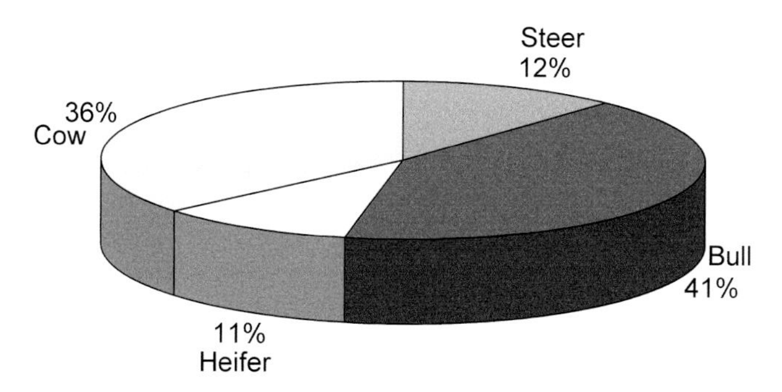

Figure 2. Distribution of cattle slaughter in Sweden, % of number.

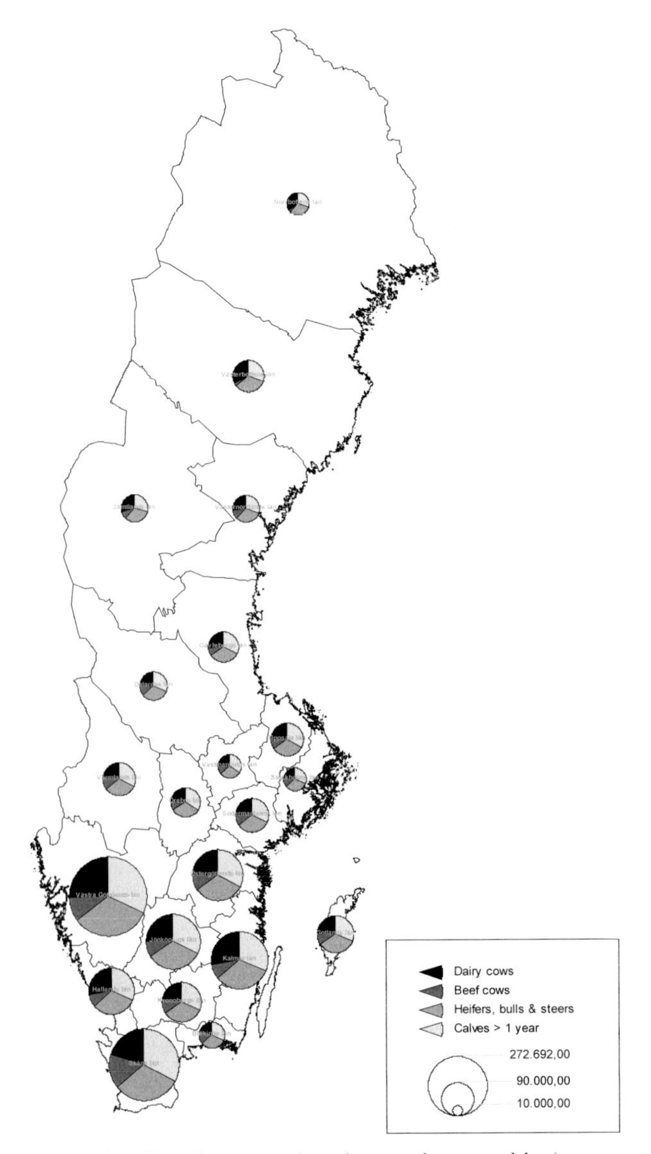

Figure 3. All cattle groups (numbers and composition).

The breeding of young stock for slaughter is either carried out by specialised breeders of young cattle or by calf producers. The feeding regime is based on silage, grain, supplementary feeds or complete concentrate feeds. In the southern parts of Sweden, corn silage is sometimes used. The use of alternative feedstuffs such as pressed beet pulp silage, potatoes, distiller's waste etc. depends on the availability of the raw material in question. The average breeding period for bulls of dairy breeds is approximately 18.9 months and for bulls of beef breeds between 17 and 18 months depending on the breed. Common housing systems for young animals are slatted floor, free stalls, straw beds and sometimes also tie stalls.

Developments in Swedish beef production

The number of beef farms has dropped considerably during the past 25 years; from somewhat more than 70,000 to 25,000. However, the reduction levelled out during the 2000's, but the downward trend continues. There are several reasons for this development, one being the age structure. In 2005, every fifth farmer was 65 years old or more. This will lead to a continuous reduction of the number of beef farms in the years to come (Figure 4).
An increasingly strained economy on the farms calls for more efficient units to enable labour wages etc. The development from 1980 to 2006 shows that farms are becoming bigger and bigger. The suckler cow herds are still very small (Figure 5). The structure of Swedish suckler cow herds is largely dependent on the availability of natural grassland and the location of the grassland. In many parts of Sweden, the grazing ground is spread out, with very small connected pieces of land.

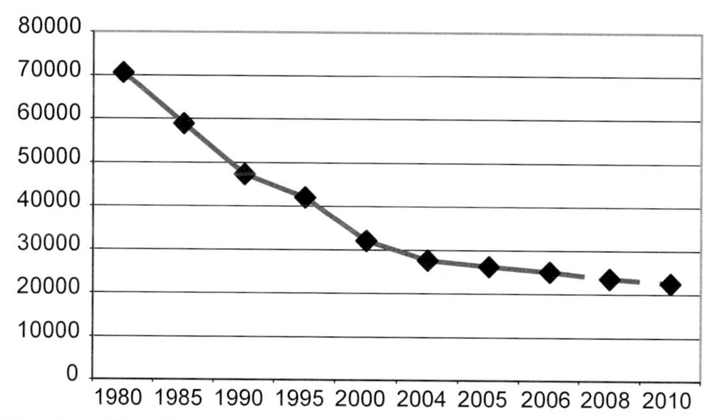

Figure 4. Number of beef farms in Sweden.

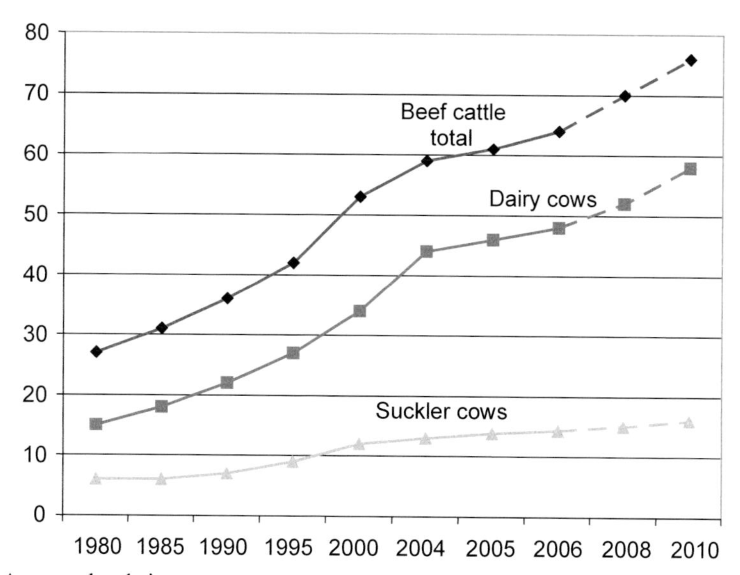

Figure 5. Average herd size.

The implementation of the Mid-Term Review reform in Sweden

On January 1, 2005, the new MTR system was implemented in Sweden. Sweden chose a mixed model comprising the single farm payment and a coupled bull premium. The coupled part of the bull premium remained at 75% of the previous bull premium (Figure 6).
The previous production-based farm subsidies (hectarage payment and animal premium) were replaced by a new support system decoupled from production, the single farm payment. To be eligible for the single farm payment, a farmer requires payment entitlements. Such payment entitlements are based on the hectare.
The bull premium quota is made up of two parts; an expense limit based on subsidies paid during the period 2000 to 2002, and an allocated quota based on the number of slaughtered male animals. The quota for male animals is not distributed on individual farms.
Modulation is made with the following percentages: 3% in 2005, 4% in 2006, 5% in 2007.
Compensatory payments are intended to support agriculture and compensate the costs of production in areas with more difficult production circumstances (LFA, Less Favoured Area) (Table 3). Such compensatory payments are paid to special support areas in northern Sweden, to certain wooded parts of southern and central Sweden, and to some areas on the islands of Öland and Gotland. In the first place, the compensatory payments are intended for hayfield and grazing land, but certain support areas in northern Sweden also receive such allowances for the cultivation of grain and potatoes. A condition for receiving compensatory payment is ownership of cattle, sheep or goats with a minimum number of animals per hectare.

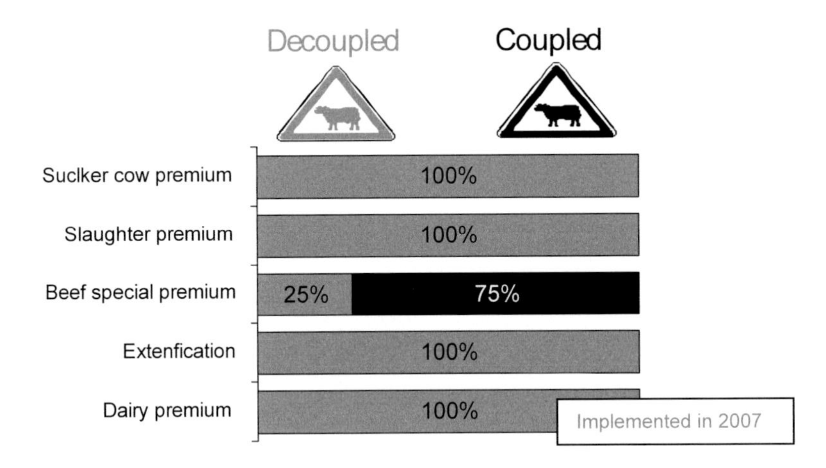

Figure 6. Decoupling scheme applied in Sweden since 2005.

Table 3. Compensatory payments in 2006 (SEK, hectares, grassland and grazing land).

Support area	0-90 ha	More than 90 ha
1 and F	2,550	1,275
2-4:a	1,950	975
4:b	950	475
5:a	1,350	675
5:b	760	380

Sweden has various types of environmental subsidies. For beef production, the most important support is the one granted for the preservation of hay meadows and grazing land. This support is intended to compensate for management costs and consists of a basic payment, which is the same for all Sweden, 1,100 SEK per hectare, plus an additional payment of 1,400 SEK per hectare for areas of high biological value and/or cultural value.
The income of Swedish suckler cow farms has drastically changed due to the complete decoupling of the suckler cow premium. In order to reach profitability in the suckler cow production, the farmers are now completely dependent on financial allowances for natural grasslands and/or organic production. For Swedish breeders of young cattle it is essential that the male animal premium is maintained at 75%. This has a positive effect on profitability and vouches for Swedish calves being in great demand.

How are beef producers going to develop their business?

According to a study carried out by the Federation of Swedish Farmers in October 2005, more than every third farmer wanted to increase the production, and this is a much more positive response than in 2004. In the group with more than 50 cattle delivered for slaughter per year, more than 50% of farmers wanted to increase their production. Every fourth farmer expressed the desire to make substantial changes. The most common spontaneous answers were that they either wanted to wind up their business or change from dairy to beef production.
The three most common incentives to increase production are to acquire more acreage, to increase the degree of occupation (higher income) and to improve profitability. In 2005, improved profitability was the main reason to increase production. On the other hand, poor profitability, together with age and shortage of land, are also some of the main reasons not to increase production.
All the above clearly show that beef producers have different views and conceptions of the beef market depending on their situation. Some essential factors are age, farm size, acreage, suitable buildings, production aim, etc.

Economic result following the Mid-Term Review reform

LRF Konsult, a company owned by the Federation of Swedish Farmers, offering financial advisory services to small enterprises, make annual evaluations of the economic result for different agricultural business. They also issue a prognosis for the year to come. The basis for this prognosis includes a database comprising 173 beef farms with data going ten years back in time. These beef farms are a mix of both full-time and part-time operations. They are made up of suckler calf producers, beef finishers as well as a combination of both. By tracing their profitability before and after the introduction of the MTR reform, a picture of the development can be formed.
As can be seen in Figure 7, profitability has had a positive development in spite of the fact that the coupled support for suckler cows, the extensification premium, the slaughter premium and 2% of the male animal premium have been removed. For 2006, the single farm payment makes up approximately 62% of the income from work and capital. During 2005, the first year after the implementation of the reform, the corresponding figure was 75%. In 2006, environmental aids in the form of grants for grazing land make up approximately 8% of the total income from work and capital. In 2007, the environmental aids will become more coupled to the number of animals, and for farms with many animals the environmental subsidies are expected to increase.
The black line in Figure 7 shows the development of the income from work and capital, not taking into account the single farm payment.

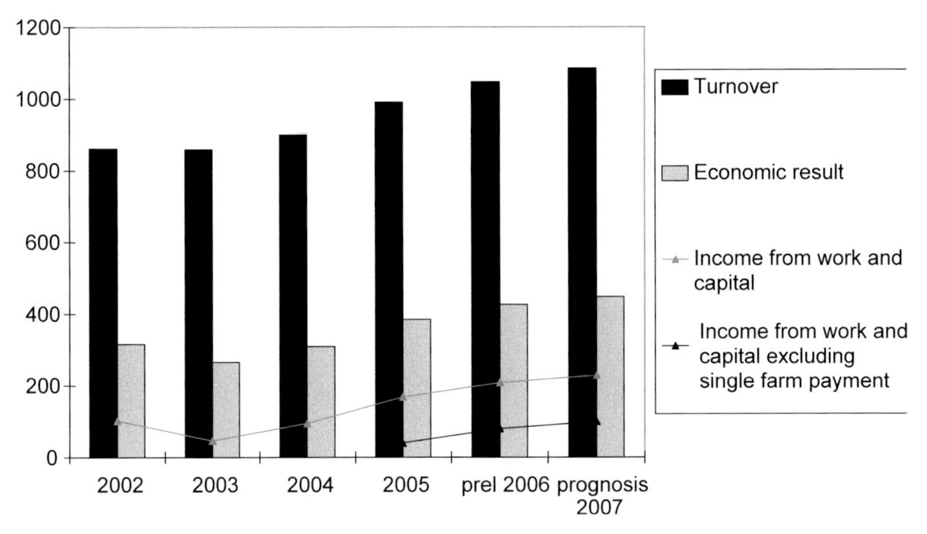

Figure 7. Economic Result, Beef Farms (thousand SEK).

Results from different types of production

The 173 beef farms were divided into three groups as is shown in Table 4.
In the beef finishers group, calves of dairy breeds make up the major part. They are slaughtered at approximately 19 months of age and the weight is then around 310 kg. The remaining young cattle are crosses with dairy breeds or pure beef breeds. The slaughter weight for this group varies between 310 and 370 kg and the age is 17-18 months.
It could be said that the groups with 1-50 suckler cows and specialised breeders of young cattle represent the herd types that are presently increasing the most. The comparison of the hourly earnings of these groups shows that without the single farm payment, the group with 1-50 suckler cows has practically no income at all from work and capital. This difference in profitability could be perhaps explained by the fact that the production-based premium for male animals still remains. The large suckler cow herds show the best profitability per hour. As farm layouts and the structure of the landscape make size rationalisation difficult in many areas with much grazing land, it is unlikely that this farm type will become common (Figure 8).
Up to now, the cost for land in Sweden has been low, but rather dramatic changes can now be seen. The price of land in good arable areas or in areas with other surplus values is four to five times higher than in areas with low alternative value. The opportunities for suckler calf production can be

Table 4. Classification of beef farms and SFP.

Farms with:	1-50 suckler cows	> 50 suckler cows	Beef finishing
Single farm payment per hectare	2,053	1,938	1,848
Acreage per farm, hectar	55	214	70
Number of working hours per farm	1,800	4,400	2,200
Number of other cattle per farm	70	247	109
Number of suckler cows per farm	19	96	0

Numbers given are mean values per farm.

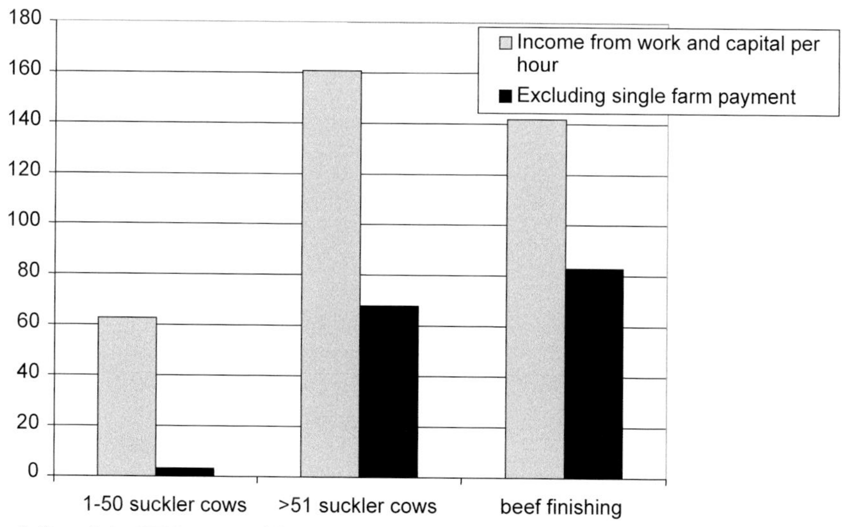

Figure 8. Result in SEK per working hour.

found in these areas with low land value. An efficient calf production together with environmental subsidies for pasture may very well generate profitable enterprises.

Specialised breeding of young cattle is connected with intense crop farming. Producing cheap feeds increases the profitability of the beef production process. Therefore, such farms show the highest profitability per hectare (Figure 9).

Figure 10 shows the size of the single farm payment and an estimated environmental aid for grazing land. As can be seen from the diagram, the income from work and capital would become negative for farms with 1-50 suckler cows if they did not receive these subsidies. Bearing in mind that these

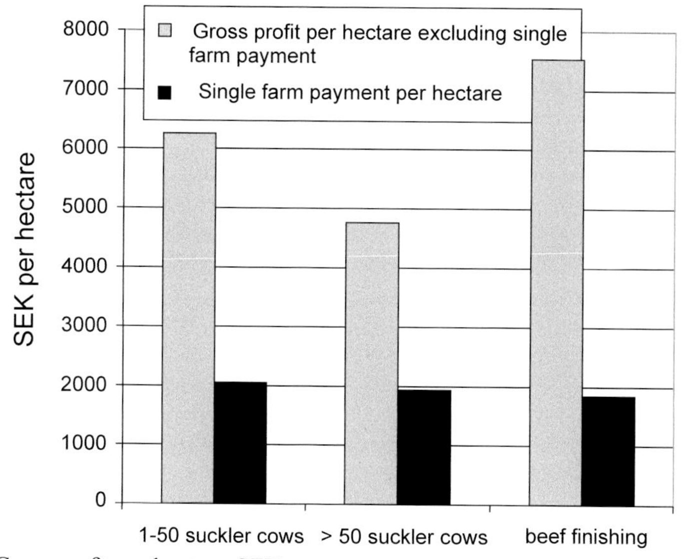

Figure 9. Gross profit per hectare, SEK.

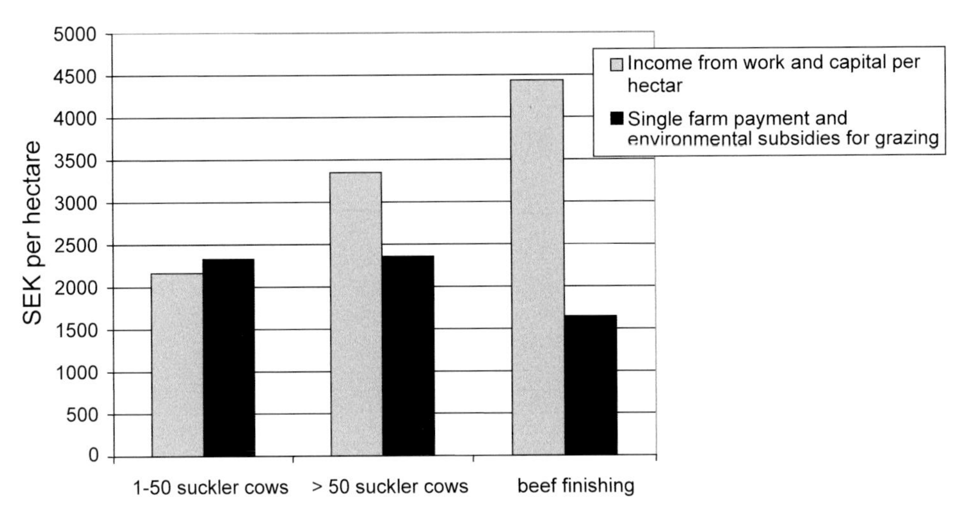

Figure 10. Total income from work and capital compared to income from single farm payment and payment for grazing.

farms often receive additional grants for organic crop farming and animal production, it is clear that this type of production is totally dependent on subsidies for its existence.
If land prices should go up, or if alternative usage of land for other purposes should develop, this would change the conditions for Swedish beef production.

Conclusion

Without any doubt, the CAP reform has contributed to the decrease of the number of beef cattle in Sweden. Just before the reform and during its implementation, the will to make capital investments was low, due to a great uncertainty as to the future conditions for beef production. The will to invest has now returned, and many beef producers have achieved improved profitability after the reform. One of the reasons is naturally the increased Swedish producer price. This increase of the producer price is not likely to continue because, in the long term, the result of the pending WTO negotiations could be seen as a threat. To a very large extent, the Swedish producer price is affected by the price of imported meat.
Another reason for improved profitability is an increased awareness of the need to rationalize production. For those beef producers who receive the decoupled premiums, the total support payment has not decreased by very large amounts. Nevertheless, these producers have shown a growing interest in optimising efficiency as well as in adjusting the production to meet market demands. Being efficient and market-oriented has become increasingly important for the Swedish beef production. This will also lead to increased specialisation.
Swedish beef production is presently encountering several threats, one being the shortage of calves. More calves need to be born in Sweden in order to secure a solid and competitive production. Certainly, in order to have more calves, the ongoing reduction of the number of dairy cows must stop. However, it is strongly believed that the number of suckler cows will still increase as a result of the constant focus on grazing land in Sweden.
The regulations on cross-compliance to be met in order to receive EU grants cause uncertainty and concern to many beef producers. There are even those who choose not to be part of the system. The

Swedish regulations on cross-compliance are tougher than those stipulated by the EU. Many people believe that this leads to impaired competitiveness for the Swedish beef producers.
The future focus on bioenergy will mainly affect the prices of feedstuffs, but to some extent also the prices of land. An estimation of future costs in Swedish beef production is therefore uncertain.

References

LRF Konsult, Nordiska Undersökningsgruppen, 2005. Swedish beef production.

LFR Konsult, 2007. Lantbrukets Lönsamhet 2006. Available online at http://www.konsult.lrf.se/data/internal/data/10/18/1170429387872/Original_uppdaterad_lantbrukets_lonsamhet_2006.ebook.pdf

Pehrson, I., 2005. New challenges for Swedish beef production. Swedish Beef Producers Association. Presentation at the 1st Cattle Network Workshop, 3-4 June 2005, Uppsala, Sweden. Available online at http://www.cattlenetwork.net/docs/workshop/presentations_pdf/Inger_Pehrson.pdf

The Swedish Board of Agriculture, 2007. Online resources. http://www.sjv.se/

Impact of the CAP reform on the evolution of beef production and beef farming systems in Europe: synthesis

P. Sarzeaud[1], *F. Bécherel*[2] *and C. Perrot*[3]

[1]*Institut de l'Elevage, Rond point Le Lannou, 35042 Rennes Cedex, France*
[2]*Institut de l'Elevage, Bd des Arcades, 87 060 Limoges Cedex 2, France*
[3]*Institut de l'Elevage, 149 Rue de Bercy 75595Paris Cedex 12, France*

Abstract

One of the objectives of the Beef Task Force of the EAAP Cattle Network Working Group was to analyse the evolution of the European beef production and beef farming systems through the last CAP Reform. The present synthesis presents first a current register of beef farming systems in Europe: evolution of cow herds, beef farming classification and main developments (enlargement and specialization). After a short presentation of the Mid-Term Review (with different options among the countries), we analyse the first consequences on European farms and then on markets and territories.

Keywords: classification, beef farms, decoupling, CAP impact.

Introduction

With 8 millions of beef TEC (carcass equivalent tonnes), the EU-25 countries contribute by 13% to the world beef and veal production and are engaged in the world market exchange. It is mainly the Common Agricultural Policy that determined two main objectives for EU farmers in the past: initially, to guarantee self-sufficiency in basic foodstuffs in response to post-war food shortages and, more recently, to reinforce the social and environmental role of the sector.
Focusing on the objectives of monitoring and analysing the beef farm strategies through the impact of the CAP, the Cattle Network's Beef Task Force compiled national studies in order to report the actual situation of the beef farming systems, few years after the Mid-Term Review application. This synthesis is structured as follows: i) the current situation of EU beef production and beef farming systems, ii) the application of the Mid-Term Review, and iii) its first consequences on farms and on their involvement in production and land use.

Current status of beef production in Europe

Evolution of cow numbers

In the twenty-year period from 1985 to 2005, beef cow numbers increased considerably in France (by 200%), Spain (by 170%) and Ireland, they rose slowly in Germany with respect to the reunification and the additional quota that was allocated, whereas they remained stable in Italy and Sweden, and decreased dramatically in eastern countries like Poland. Most of these changes occurred before 2000 and can be linked to the implementation of the CAP policies in force at that moment.
Between 2000 and 2005, the overall number of suckler cows increased slightly in Europe although it decreased in Italy and Germany, but it remained stable in France, Spain and Ireland and increased in Sweden.

Over the same period, dairy cow numbers decreased more or less in every country because of the milk quota policy of the 1980's and the implementation of breeding programs aimed primarily at increasing milk production per cow (Table 1). Initially, these programs were more efficient in the northern part of Europe but gradually they were adopted more widely. At present, the replacement of dairy farms by beef farms is occurring in a number of countries.

Table 1. Dairy and suckler cow numbers in some EU countries and overall.

000 head	Dairy cows				Suckler cows			
	2001	2005	2006	06/01	2001	2005	2006	06/01
Germany	4710	4164	4030	- 14.5%	804	732	731	- 9.1%
Spain	1207	1018	980	- 18.8%	1895	1954	1960	+ 3.5%
France	4424	3895	3800	- 14.1%	4218	4029	4075	- 3.5%
Ireland	1174	1101	1086	- 7.5%	1160	1150	1150	=
Italy	2126	1842	1840	- 13.4%	443	472	468	+ 5.6%
Sweden	447	391	378	- 15.4%	158	164	167	+ 5.7%
EU 15	21024	18438	18048	- 14.1%	11878	11756	11743	-1.1%
2004 New Members	4949	4543	4476	- 9.5%	272	325	363	+ 33.5%
EU 25	25973	22981	22524	- 13.2%	12150	12051	12106	=

Source: Breeding economical report N° 365, French Livestock Breeding Institute, February 2007

Beef production comes mainly from the dairy herd in Germany (88%) and Sweden (70%), but it is shared more equally between the dairy and beef herds in France (40%) and Ireland (50%). In Italy and Spain, over 70% of final beef production comes from the beef herd sector due to the existence of specialised beef farms finishing imported weaners (45% of total production).

Classification of beef farms

The 2004 FADN has a new classification of beef farming systems showing that there are 491,000 beef producers in the EU-25. Of these, 58% are cow-calf farms, 17% practice fattening without rearing, and 25% are mixed dairy and beef producers. This diversity of beef farming systems is due to the adaptation of the livestock management to forage availability (grass or crops) and to market opportunities.
Within the overall mix of systems, two types of beef production need to be distinguished:

i. The rearing and finishing of beef 'on the farm' (or farms with a complete cycle) with the production of steers or heifers on grassland (e.g. in Ireland and UK) or bull fattening with maize silage (e.g. in France and Germany).
ii. The production of bulls or heifers born in cow-calf farms, mainly located in the mountains of France and the Mediterranean areas of Spain and Italy, and fattened in Italian and Spanish feed lots with forage crops and concentrates.

Increasing size and specialisation of beef farms

The size of beef farms is generally increasing in all the western EU countries (while it varies for eastern countries) but the speed of this evolution and the reasons for it differ between countries. In Ireland, for example, where beef farms are relatively small compared with some other countries in

Europe, size is increasing very slowly (by about 0.6 ha per year) because of low land mobility and high land prices.
On the other hand, in countries like France and Sweden the size of beef farms has increased rapidly in recent years, averaging 2% to 3% per year. In France, land availability is facilitated by the retirement of farmers and by relatively low land prices. This situation also applies to Sweden where suckler herds still remain small (less than 15 cows per herd in 2005) although dairy farms are much larger.
In Spain, the MAPA (Ministry of Agriculture) highlights the remarkable growth of the beef sector, due to both the support by the CAP regulations and the ability of the sector to improve competitiveness and be able to face the strong competition from other countries, thanks to technological improvements and specialization.
Some projections have been made for both Ireland and France in 2015. By then, the total number of cattle farms (both dairy and beef) is projected to decrease by 20% in Ireland and by 30% in France, accounting for 20% of the national herd. Dairy farm numbers will decrease more rapidly than beef farms in both countries. The relatively high proportion of farmers over 60 years of age is also an important consideration in beef farming.
In Germany and Italy, where fattening systems are prevalent, the small fatteners are exiting to the benefit of the larger ones. Such exodus has been reinforced by the implementation of Agenda 2000 and the two BSE crises (1996 and 2001).

The Mid-Term Review and the application of decoupling

The Mid-Term Review (MTR) reform that is being implemented since 2005 is responsible for important modifications to the European Agricultural Policy. The first and most important element of these modifications is the decoupling of the direct payments from production. The intention is that payments are made directly to farmers through a Single Farm Payment (SFP), based on historic references, and no longer linked to agricultural production.
The second element is that the application of the decoupling reform is not uniform across member states. Each country was allowed to choose its own process of implementation and the year of commencement.
The third element is cross compliance. This entails the linkage of payments to the compliance with environmental regulations. The SFP is modulated in order to consolidate farms in development and this is effectively the first impact of the reform on the level of payments. Cross compliance consists in gradually reducing the payments from the first pillar and allocating them to the second pillar measures, basis of a rural and structural policy and environmental programmes.
By 2006, while the majority of European beef production was totally decoupled, about one third still remained coupled retaining the suckler cow premium (more than half of the beef cows), slaughter premium or special male premium. Furthermore, for the SFP, countries such as Germany combined historic individual payments on regional basis.
In summary, the evolution thus far shows a general growth of beef farms and many national studies in the individual countries show that the implementation of the MTR reform will not change this trend.

First consequences of the CAP reform on the farms

Consequences for the farmers

To offset the reduction in farm income due to the modulation of CAP payments (5% in France), farmers – especially beef farmers – are encouraged to increase their herd size. Such a policy was confirmed by a national poll in France in 2005, organised by the Livestock Institute. It showed that

the growth in herd size should maintain returns for about 50% of the beef farmers, but 30% would have decreased returns.
A study by FAL (Federal Agricultural Research Centre) on intensive bull finishers in Germany showed that, under the German implementation framework, moderate growth would be the most profitable and least risky option up to 2013. In Sweden, a survey by the Federation of Swedish Farmers showed that amongst those that produced more than 50 cattle, over 50% intended to increase production.
On the contrary, in 2004, surveys conducted in association with the National Farm Survey in Ireland (Teagasc) showed that there were more cattle farmers who intended to decrease than increase their numbers of livestock units. In addition, 30% of dairy farmers intended to reduce or quit and 21% intended to give up dairying for beef production.
In both Italy and Spain, the objective of professional fatteners in recent years has been to enlarge their enterprises in order to be more competitive and responsive to market conditions.
According to a CRPA (Research Centre for Animal Production) analysis in Italy, the first reaction of the beef farmers to the CAP reform was an attitude of 'wait and see'. In the first months of 2005, when decoupling of premia was introduced, many large fattening farms maintained their stables empty in order to verify the new market conditions. Some small farms decided to cease production. This reaction to the decoupling of premia caused a reduction of national beef production of 6% in 2005.

Impact of the new CAP reform on economic results

2005 was the first full year of total decoupling in Ireland, Germany, Italy and Sweden. In those countries, the future profitability of beef finishing and cow-calf production depends on the profitability of production and not on direct subsidies. In 2005 in Ireland, the income for cattle farming increased by 35% to 40% according to the NFS (National Farm Survey), due largely to carry over of direct payments from 2004. When the effects of the 2004 direct payments were discounted, the increase was just 10% while cattle prices increased by 5%.
In Germany, according to FAL, the economic situation on whole farm level should have improved on many farms because of better beef prices and constant single farm payments. This does not mean that producing beef and keeping suckler cows is profitable.
In Sweden, the analysis conducted by LRF Konsult shows that the income from work and capital, excluding the single farm payment, of 173 beef farms (suckler calf producers and beef finishers) should increase in 2006 and probably again in 2007.
In Italy, sale prices increased considerably (+11%) due to a fall of production thus resulting in better profitability for the beef finishing farms. In 2005, the total return (sale prices + single farm payments) covered 113% of total cost against only 105% in 2004. In France, the CAP reform was implemented in 2006. In 2005, according to the French Livestock Institute, the incomes of the suckler cow producers and young bull fatteners were comparable to those of 2004 due to similar increases in both output and costs.
In 2006, because of the increase in the cattle prices, the incomes of beef farmers should increase everywhere in Europe. But in 2007, prices are again decreasing everywhere in Europe, particularly for young bulls in France and Italy. That may be the end of the improvement in the beef market, which has been continuous since the European crisis of 2001. Over the coming years, except for that part of the production that is not fully decoupled (suckler cow and part of slaughter premium in France and Spain and slaughter premium in the Netherlands as well as the special premium in the Scandinavian countries), beef production and especially fattening production will be directly linked to the market situation.

Compensatory payments and subsidies for pasture can influence the income of the cow-calf farms as explained by Taurus in Sweden (in 2006 such environmental aids represented 8% of the total income of work and capital of the farms sample).
In France, the Government chose to maintain the grassland premium that supports mainly the cow-calf producers in the grassland areas. Because of the dynamic 'hybrid decoupling model' adopted in Germany, the grassland payments there will gradually increase after 2009 to reach the same level of the cropland payments while the single farm payments will be phased out. By favouring grassland, the choice could weaken the activity of beef finishing in intensive livestock farms, but the delay to full implementation should allow these farms to adjust to the new regime (FAL).

Consequences for cow-calf producers

In countries where the suckler cow premium remains coupled (e.g. France and Spain) and taking account of the role of beef production in the economy of grassland regions, the number of suckler cows should remain stable until 2013, provided that the fattening activity in the South of Europe continues as at present.
Taking account of the increase in the size of the farms (see above), the French cow-calf producers will have to choose between the production of young weaners (from the Blonde d'Aquitaine or Limousine breeds) intended for supply to Spanish, French and certain Italian fatteners, and heavy weaners (Charolais breed) for supply to the Venetian fatteners. At present, the cow-calf producers are ready to adapt their production (type of weaners, season of production, calving period, and guarantees of production) to the demands of the importers. They will be helped in this by the decoupling of the special premium previously attached to backgrounders.
In Spain, the weaning activity in the western or central areas of the country is directly linked to the fattening activity in the north-eastern or central areas. That production will be sustained for as long as the fattening activity is profitable. In fact, the weaning prices remain related to the European beef market and especially the buoyancy of beef consumption in the South of Europe.
Irish suckler beef systems that export weaners are also ready to adapt insofar as the economic opportunities permit. Finally, with the environmental subsidies the number of suckler cows could increase in Sweden. In Spain, from 2006 the additional aids (article 69) for the extensive holdings concern mainly the suckler cow herds in the centre-west of the country. That might help to maintain the number of suckler cows that is one of the main objectives, in particular to secure the provision of the Spanish fatteners. Another aim is to improve the fertility of the suckler cows by securing the feeding in the drought areas.
On the contrary, according to CRPA, the profitability of suckler cow farms increased in 2005 in Italy because of a rise in sale prices and a reduction of production costs, but there is no guarantee for such a level of prices in the future.

Consequences for fatteners

In Italy, in 2005 the implementation of the CAP reform did not have a significant impact because of the market context and the beneficial effects of Avian Influenza on Italian beef consumption. The increase in bull prices resulted in improved profitability independent of the decoupled payment. This situation was further helped by the relatively constant European production due to the stabilisation of suckler cow herds. Ceasing of production and reliance on the SFP by small beef fatteners (under 100 or 200 bulls) was also expected but this seems to have interested few people.
Most of the beef fattening farms are highly specialised in beef production. Improving profitability gets through the search of the highest independence possible of the feed market. For this reason,

the decoupling of premia does not open a strategy of diversification, as a high rate of specialisation remains the main objective of fattening farms.
The increase of environmental regulations and the implementation of the Nitrates Directive in the Po Valley from 2007 could be a new constraint for Italian beef production, especially for small producers, but it could also improve the competitiveness of the professional operators. The large farms will continue to enjoy comfortable Single Farm Payments that will enable them to face the Nitrate Directive problem with less financial efforts than the small farms, which could disappear. However, eventual decisions about maximum ceilings of CAP payments after 2008 might affect these large farms in a worse way.
In Spain, the fattening sector in the future needs to be provided sufficiently and at a good price by weaners coming from Spanish or French suckler cow farms or by calves coming from Poland. According to the MAPA, in the next years the implementation of CAP in Spain should give a proper answer for the inside market, especially on a beef quality differentiated market, but there is uncertainty regarding the supply of foreign weaners or calves in connection with the 'bluetongue disease', a possible development of fatteners in Poland and new welfare constraints (transport of live animals).
In France and Germany, the system of mixed cow-calf (or dairy) and fattening enterprises is particularly affected by decoupling. In the coming years, bull finishers can more easily decide to cease, reduce or increase their activity without losing direct payments. In 2005 and 2006, the balance between the sale price of bulls and the purchase price of the weaners was in favour of fattening, compared to crop production, but the balance has changed in 2007. Crop prices have now risen in response to the recent increase in demand for energy crops, firstly in America but also in Europe. Teagasc underlines the increasing competition between animal production and bio-fuel for the world supply of grain. Because of the increased use of grain in Irish beef fattening in recent years, if the price of grain increases the margin in beef production will decline, obliging beef fatteners to develop production systems based more on grass and with less grain. This will have two negative consequences, namely increased methane emissions and increased seasonality of production. The Nitrates directive and associated environmental regulations could also limit the development of large-scale winter fattening units in Ireland.
In Spain, where there are major constraints related to agro-climatic conditions (deep droughts), costs of feeding based on cereal concentrates could increase and then force the producers to find new models of production. MAPA underlines that integrated systems with a high level of management on the purchases could be a solution. They also promote a vertical integration all along the sector.

First consequences of the CAP reform on markets and territories

More specialisation of beef production in Europe?

Compared with dairy, sheep or crop production, the specialisation of beef production could be reinforced by structural farm evolution. Less dairy cows and less dairy farmers should reinforce the specialisation of the beef production from suckler herds in Europe, mainly in countries such as France, Spain and Sweden.
In Ireland, according to FAPRI Ireland, up to 35% of dairy farmers could exit dairying between 2002 and 2015. But most of those farmers would engage in part-time beef farming and keep the land rather than change their production system or sell their properties. Accordingly, the percentage of part-time farmers is likely to increase as long as the economy remains positive.
In the context of the world market for cereals and sheep meat, farmers must also specialise in these enterprises. In countries where the special beef and slaughter premiums have been fully decoupled

(e.g. Italy, Germany), systems based on large fattening farms will depend on future profitability, which in turn will be a function of the price of beef meat.
The reorganisation of the beef production is necessary relatively to the competition engaged on the world market. Some costs of production could be reduced through economies of scale.
According to national analyses, that trend could contribute to maintain the production of meat from the beef herd, but it shall not be sufficient to maintain the global EU meat production because of the decline of the dairy herds, as far as genetic improvement increases milk productivity. That will have deep consequences on beef post-production activities such as slaughterhouses and carving workshops, whereas the stores could buy meat abroad.

Land use: a major issue in Europe

The impact of changing beef farming systems on land use depends both on the potential alternative uses of the land and on its cost. It could be particularly important in countries such as Ireland where part-time beef farming is common, but in such grassland areas beef production is well adapted to the maintenance of countryside at a marginal cost.
In the grassland areas of France, and also on the central plateaux of Spain, the decision to maintain the suckler premiums coupled continues as a support to cow-calf extensive production. In those areas, generally disadvantaged or mountainous, there is no alternative except forestry or sheep production, as fattening heifers or young bulls and producing steers would be less profitable than before. Suckler cows therefore contribute to maintaining populations in the countryside.
Moreover, in Spain the implementation of article 69 concerning the support of extensive holdings (< 1.5 LU / ha) should help to maintain livestock activity in those regions.
The introduction of the grassland premium in Germany and the increase of this premium to the same level as cropland premiums (around 300 EUR per ha) in 2013 is an indirect – and nevertheless decoupled – support for grassland systems. The decision to continue suckler-cow production firstly depends on the alternatives available to fulfil the cross-compliance requirements. Mulching the land is the new reference system for all agricultural activities. But if livestock production costs more per ha than mulching, it is not the profitable alternative. Moreover, the introduction of the grassland payments will lead to an increase of rental prices, thereby diminishing whole-farm profitability, especially in Eastern Germany, where the share of rented land is more than 90%. In Sweden, animal production from pasture is directly supported by environmental subsidies linked to a minimum animal density. These policies could reinforce the specialisation of areas in cow-calf production or fattening enterprises.
In more intensive areas where beef production competes with crops, competitiveness will be a major challenge for beef producers in the future. The challenge will be to ensure profitability in the whole production chain taking account of calf prices, feeding cost and beef price changes, together with increasing or decreasing levels of beef imports.

Conclusion

One of the consequences of the CAP implementation in the last 20 years has been the creation of a European beef market with more and more relations between areas of calving and areas of fattening inside and between the countries.
Whatever the choices for the implementation of the CAP, the European beef farming sector should have a similar evolution: specialization and increase of the size of beef farms linked to the disappearance of the smaller, less specialized and less competitive farms. Everywhere, fatteners will have to reduce their cost of production (feeding) in order to be able to face the competition with the foreign countries, which export more and more meat to Europe.

The countries where the suckler cow premium remains coupled and the countries that encourage the grassland areas by environmental support measures, should maintain their national herds provided that those farmers will find European (inside or abroad) fatteners to buy the weaners or the calves produced in Europe. That will not be enough to provide the market. On the contrary, the lack of calves born in Europe first from the dairy herd but also occasionally from the suckler herd represents a danger for that sector in the future.

The demand of European consumers for food security, quality of meat, respect of the environment and animal welfare, the importance of beef production for land use and human activity in so many disadvantaged areas, and the linkages between different European countries must be kept in mind when making new proposals for the CAP.

Conclusions and recommendations of the 3rd EAAP Cattle Network Workshop

Participants at the Workshop agreed that beef production, as one of the most important agricultural sectors in the European Union, attracted particular attention and support. This situation has changed with the implementation of the CAP reform. Animal science and economic research are in charge of analysing challenges and opportunities for beef production under the new circumstances created not only by the CAP reform but also by developments in international markets and in other beef producing regions, also taking into account the growing global demand for animal products, including beef.
In this respect, the Workshop commended the EAAP Cattle Network's Beef Task Force for the excellent study material on the impact of the CAP reform on the beef sector in the European Union.
The Workshop requested the EAAP Council to ensure that proceedings of the Workshop are published in the EAAP Technical Series, distributed to all stakeholders in the beef sector in the EU Member States and handed over by the official representatives of the EAAP to the relevant services of the European Commission as an EAAP contribution to the mid-term review of the CAP. The same publication should be forwarded also to COPA-COGECA. The EAAP Secretariat was also requested to place the presentations of the workshop on the website of the Cattle Network (http://www.cattlenetwork.net).
In discussing the future programmes of the Cattle Network with regard to the beef sector, the Workshop recommended that future actions should focus on the following issues:

Implementation of CAP rules and regulations and their effects on the cattle sector

a. The workshop agreed that the Cattle Network should continue to monitor developments in the CAP and initiate studies and analyses of the impact of policy changes and reforms on the cattle sector in the European Union. These studies and analyses should be brought to the attention of stakeholders and policy makers.
b. Findings and recommendations of the Cattle Network should be forwarded to the EAAP Council and to the relevant EAAP Study Commissions with a view to initiating further in-depth research and studies in the fields of cattle genetics, management and health, nutrition and production systems, thus fulfilling the EAAP mission and providing support to the adjustment of the cattle sector to new circumstances.
c. There was agreement on the necessity of further analysis of CAP effects on major beef farming systems in specific countries (following the classification of the Beef Task Force) with the aim to improve the understanding of major systems and their dynamic modification in relation to the CAP. The analysis could also be enlarged to other members of the European Union and especially new members.
   It was felt that this analysis should be based on a standardized/harmonized methodology using already existing frames for farming system analysis on national and international farm networks. It could be supported by PRA (Participatory Rural Appraisal) methods to link with stakeholder perception and assessment. Information should be used for modelling effect scenarios.
d. The workshop also addressed the need to analyse
   - interacting rules and regulations of the CAP, such as landscape/environment protection, rural development;
   - potential effects of global warming/climate change;
   - impact of price fluctuation of major feeds for beef production.

Sustainability of beef farming systems

The Workshop discussed this important issue at three levels:

a. Global
   - Analysis of trends in the global and regional markets for beef and beef products which should include production and consumption trends, consumer preferences and their respective influencing factors. This analysis needs to be connected to dairy production and market trends due to the obvious convergence between the two sub-sectors. The Workshop was aware of respective expert groups of market economists doing these studies (FAL, Germany; Livestock Institute, France; CRPA, Italy; Rural Economy Research Centre of Teagasc, Ireland). Consequences of global market changes on EU production conditions for beef are also investigated by respective expert groups (for example the agri benchmark network).
   - Further studies on the effect of bio-energy production on land use and feed prices and on the effect of internalizing the methane factor as an environmental cost in BFS are required to forecast the competitiveness and perspectives for major beef production systems. The workshop was informed that researchers of the agri benchmark network have started working on these issues.

b. Regional
   - Further information are required to measure the effects of CAP measures supporting rural development through BFS. The main focus should be put on describing sustainable cattle enterprises fitting the ecological and economic environment and on measuring their impact on rural development.
     The group also touched on the issue that it may not seem always sensible to promote beef production if the social objectives are not met.

c. Farm and Product Value chain.
   The workshop discussed a number of issues related to farm and product value chain:
   - Analysis of alternative land use scenarios and their impact on the relative economic competitiveness of cattle farming. Standardized methods for farm-enterprise analysis are available and will be included in a Session of the EAAP Study Commission on Livestock Farming Systems during the 2008 EAAP Annual Meeting in Vilnius.
   - Analysis of complete beef value chains to identify strengths and weaknesses of the chain process, of respective stakeholders and to measure transaction costs in order to improve effectiveness.
   - Analysis of lessons learned, strengths and weakness of different labelling systems, their organisational requirements and impact for beef farming systems.
   - Explore possibilities in reducing current costs in implementing the EU cattle identification system. Multi-country comparisons of options and steps discussed and pursued might be useful.

Typology of beef farming systems

The workshop agreed on the need for continuous work on the typology of beef farms in the EU Member States and on the analysis of strategies in relation to the environmental sustainability of theses systems.

a. The typology needs to be enlarged to include the new European members' specificities as far as their beef production, farm size and European livestock zoning are concerned.
b. A specific description of organisational forms of beef farming is required for various countries to contribute to an EU-wide assessment. A new step could be to specify the beef management implemented in the beef farming systems (husbandry, feeding, grassland use, etc) to compare the

main European beef productions and produce an analysis of their main strengths and weaknesses. This description will integrate methodological assessments developed in the agri benchmark international panel. The overall BFS analysis should follow procedures and methods used in a Product Value Chain approach by specifying the main products of each BFS.

c. The Cattle Network should try to link up more with the EAAP Contact Group for CEEC.

The EAAP Technical Series so far contains the following publications:

- **No. 1. Protein feed for animal production**
  With special reference to Central and Eastern Europe
  edited by C. Février, A. Aumaitre, F. Habe, T. Vares and M. Zjalic
  ISBN 978-90-76998-03-9 – 2001 – 184 pages – € 35 – US$ 39

- **No. 2. Livestock breeding and service organisations**
  With special reference to CEE countries
  edited by J. Boyazoglu, J. Hodges, M. Zjalic and P. Rafai
  ISBN 978-90-76998-04-6 – 2002 – 75 pages – € 25 – US$ 30

- **No. 3. Livestock Farming Systems in Central and Eastern Europe**
  edited by A. Gibon and S. Mihina
  ISBN 978-90-76998-29-9 – 2003 – 264 pages – € 39 – US$ 51

- **No. 4. Image of the Cattle Sector and its Products**
  Role of Breeders Association
  edited by J. Boyazoglu
  ISBN 978-90-76998-33-6 – 2003 – 88 pages – € 27 – US$ 32

- **No. 5. Foot and Mouth Disease**
  New values, innovative research agenda's and policies
  A.J. van der Zijpp, M.J.E. Braker, C.H.A.M. Eilers, H. Kieft, T.A. Vogelzang and S.J. Oosting
  ISBN 978-90-76998-27-5 – 2004 – 80 pages – € 25 – US$ 30

- **No. 6. Working animals in agriculture and transport**
  A collection of some current research and development observations
  edited by R.A. Pearson, P. Lhoste, M. Saastamoinen and W. Martin-Rosset
  ISBN 978-90-76998-25-1 – 2003 – 208 pages – € 40 – US$ 53

- **No. 7. Interactions between climate and animal production**
  edited by N. Lacetera, U. Bernabucci, H.H. Khalifa, B. Ronchi and A. Nardone
  ISBN 978-90-76998-26-8 – 2003 – 128 pages – € 35 – US$ 39

- **No. 8. Farm management and extension needs in Central and Eastern European countries under the EU milk quota system**
  edited by A. Kuipers, M. klopcic and A.Svitojus
  ISBN 978-90-76998-92-3 – 2006 – 278 pages – € 44 – US$ 59